Conference Board of the Mathematical Sciences
REGIONAL CONFERENCE SERIES IN MATHEMATICS

supported by the
National Science Foundation

Number 23

LECTURES ON SET THEORETIC TOPOLOGY

by

MARY ELLEN RUDIN

Published for the
Conference Board of the Mathematical Sciences
by the
American Mathematical Society
Providence, Rhode Island

Expository Lectures
from the CBMS Regional Conference
held at the University of Wyoming
August 12–16, 1974

AMS (MOS) 1969 subject classifications: Primary 0265, 5420.

Library of Congress Cataloging in Publication Data CIP

Rudin, Mary Ellen, 1924–
Lectures on set theoretic topology.
(Regional conference series in mathematics; no. 23)
"Expository lectures from the CBMS regional conference held at the University of Wyoming, August 12–16, 1974."
Bibliography: p.
1. Topology–Addresses, essays, lectures. 2. Set theory–Addresses, essays, lectures. I. Conference Board of the Mathematical Sciences. II. Title. III. Series.
QA1.R33 no. 23 [QA611.15] 510'.8s [514] 74-31124
ISBN 0-8218-1673-X

Printed in the United States of America

CONTENTS

Other Monographs in this Series

No. 1. Irving Kaplansky: *Algebraic and analytic aspects of operator algebras*
2. Gilbert Baumslag: *Lecture notes on nilpotent groups*
3. Lawrence Markus: *Lectures in differentiable dynamics*
4. H.S.M. Coxeter: *Twisted honeycombs*
5. George W. Whitehead: *Recent advances in homotopy theory*
6. Walter Rudin: *Lectures on the edge-of-the-wedge theorem*
7. Yozo Matsushima: *Holomorphic vector fields on compact Kähler manifolds*
8. Peter Hilton: *Lectures in homological algebra*
9. I. N. Herstein: *Notes from a ring theory conference*
10. Branko Grünbaum: *Arrangements and spreads*
11. Irving Glicksberg: *Recent results on function algebras*
12. Barbara L. Osofsky: *Homological dimensions of modules*
13. Michael Rabin: *Automata on infinite objects and Church's problem*
14. Sigurdur Helgason: *Analysis on Lie groups and homogeneous spaces*
15. R. G. Douglas: *Banach algebra techniques in the theory of Toeplitz operators*
16. Joseph L. Taylor: *Measure algebras*
17. Louis Nirenberg: *Lectures on linear partial differential equations*
18. Avner Friedman: *Differential games*
19. Béla Sz.-Nagy: *Unitary dilations of Hilbert space operators and related topics*
20. Hyman Bass: *Introduction to some methods of algebraic K-theory*
21. Wilhelm Stoll: *Holomorphic functions of finite order in several complex variables*
22. O. T. O'Meara: *Lectures on linear groups*

I. Cardinal Functions in Topology

One hundred years ago [1] Cantor introduced the concept of one-to-one correspondences as a way of measuring the size of infinite sets. Consequences of this concept included the immediate birth of the continuum hypothesis (henceforth known as CH), set theory, and topology. Today topology and set theory are diverse and complicated subjects. Set theory uses the language of mathematical logic which has developed a precise and complex structure which is not part of the background of the average mathematician who is not a logician. Topology ranges through global analysis, algebraic topology, and geometric topology, to a coverall for the multitude of remaining topics called general topology. It is certainly true that the study of metric manifolds with and without differential or algebraic structure is the most important function of topology and the one which makes it vital to the rest of mathematics; the tools are analytic, algebraic, geometric, countable and finite combinatorics; and most topologists are engaged in this study. However a smaller group, and I am one of these, is primarily concerned with abstract spaces and in this area a real revolution has been going on in the last ten years because of renewed interaction with set theory. One should stress that not all of the current activity in general topology is a result of this more set theoretic way of thinking: the beautiful work in infinite-dimensional spaces and shape theory take their tools primarily from manifold theory in general and gemoetric topology in particular.

But the set theoretic influence on general topology is an exciting, effective change and it is this change which this series of lectures will study. The range of spaces considered will be primarily from Hausdorff to compact or metric. We will assume throughout that all spaces are Hausdorff unless its absence is specifically mentioned. If a space is both compact and metric then it has a countable basis for its topology and most of the properties which concern us will be trivialized. However, we will be concerned with metrizability and compactifications. Still, the area which will command the largest attention is that large hole between regularity and normality on the one hand, and compactness and metricity on the other.

From the time of Cantor until 1920, topology and set theory were almost indistinguishable. Each was primarily concerned with axiomatics, with understanding the real numbers, with making the proper definitions. *Fundamenta*, a journal primarily devoted to this area, was founded in 1920 and the following ten years was the real golden age of general topology. The most fundamental theorems, those which are part of every mathematician's background, were proved at this time. For example:

[2] TIETZE'S EXTENSION THEOREM AS REFINED BY URYSOHN. *A space is normal if and only if, for every closed subset K of X and continuous function f from K into the reals, there is a continuous extension of f to X.*

[3] ALEXANDROFF-URYSOHN METRIZATION THEOREM. *Every regular space with a countable basis is metrizable.*

[4] THE TYCHONOFF THEOREM. *Every product of compact spaces is compact.*

These theorems illustrate the virtues and blind spots of the 1920's. Since extending continuous real valued functions is of the very essence of mathematics, Tietze's extension theorem shows that the class of normal spaces is one of the most fundamental, but mathematicians of the 1920's had few clues as to how to work with normal spaces. They searched in vain for a reasonable class of spaces to fill the gap between normality and metricity or compactness.

The metrization theorem is beautiful, useful, the only metrization theorem most mathematicians care about. It says that a countable basis for a regular space destroys all of the set theoretic pathology. But the theorem does not tell us of the pathology which may be present in a metric space. The theorem illustrates the tendency of the 1920's topologist to think in terms of cardinalities as being simply countable or uncountable. This fact often led him to beautiful consequences of the countable case but also to an inability to analyze the uncountable case.

The Tychonoff theorem does not suffer from any lack of generality! It is totally beautiful and illustrates the fact that, even through the 1920's, topology and set theory moved hand in hand. This is a theorem in set theory perhaps even more than in topology. It also illustrates the parts of set theory used by 1920's topologists: they felt quite at home using the axiom of choice, they understood and used elementary cardinal arithmetic, and they were aware of CH and its many consequences. Proofs tended to be linear: by well-ordered sequences.

We shall illustrate at length that tree type proofs have been used quite effectively in the 1970's to solve many simple important problems posed in the 1920's. However, one should emphasize that the insight of the 1920's topologist led him to pose many elegant and still difficult problems. I am reminded of Cantor's opinion that the art of properly posing a mathematical problem is sometimes more important than solving it.

In the 1930's Gödel proved two well-known theorems.

[5] *CH is consistent with the usual axioms of set theory.*

[6] GÖDEL'S INCOMPLETENESS THEOREM. *If T is a consistent axiomatized system including arithmetic, then there is a closed formula A in T which is undecidable in T.*

One consequence of the incompleteness theorem was that set theory and topology began to move in different directions. The set theorist recognized that his subject is much more complex and philosophical than had been previously understood; he began to move away from simple cardinal arithmetic into the linguistic substructure. The topologist on the other hand decided this was a trip he did not want to take. Topology in the 1930's began to turn away from set theory to the study of manifolds, particularly to algebraic topology.

In the late 1940's and early 1950's, general topology had a strong upsurge brought on by the definition of paracompactness [7]. One easily proves that paracompact spaces are normal and compact spaces are paracompact. Stone proved [8] that metric spaces are paracompact. Paracompact spaces proved to be the "right" class between normal and compact and metric spaces. Another class developed at this time and lying between normal and paracompact spaces is that of collectionwise normal spaces [9]. The concept of studying abstract spaces in terms of discrete or locally finite families of open sets really opened our eyes to the set theoretic pathological possibilities. Some old problems were quickly solved; others were translated into set theoretic terms. The tools were still those of the 1920's; but the recognition of the necessity for distinguishing uncountable cases led to many theorems and examples.

The most beautiful and important is surely:

[9], [10], [11] SMIRNOV-NAGATA-BING METRIZATION THEOREM. *A space is metrizable if and only if every open cover has a σ-locally finite (Bing, σ-discrete) basis.*

Many familiar counterexamples came out of this period: the *Sorgenfrey line* [12], the *Michael line* [16], *Bing's G* [9]. Other favorites of mine include:

[17] *Dowker's* proof that $I \times Y$ (I is the closed unit interval) *is normal if and only if* Y *is normal and countably paracompact.*

[9] *Bing's* proof that *a Moore space is metrizable if and only if it is collectionwise normal.*

[13], [14], [15] *Michael's* beautiful analysis of paracompactness.

Compactifications also prospered at this time by being looked at in more set theoretic terms. Gillman and Jerison's *Rings of continuous functions* [20] unified this important development and Hewitt's original paper [21] is still worth knowing.

Geometric topology was really the dominant new topological theme in the 1950's and differential topology in the 1960's. Algebraic topology did not take a back seat in either development. But something happened in the 1960's which had profound effect upon the part of topology we are concerned with.

[19] *Paul Cohen* proved that *it is consistent with the usual axioms for set theory that the continuum hypothesis be false.*

In itself this theorem has few consequences in topology for there is very little one can do with not-CH alone. But the technique of proof, called forcing, has translations into Boolean algebra terms, into partial order terms, into terms which lead to remarkable combinatorial statements which are applicable to a wide variety of topological problems related to abstract spaces.

By the late 1960's there was tremendous activity in general topology brought on by the combination of the abundance of set theoretic topological questions, a new awareness of a variety of combinatorial techniques, the new translations of forcing techniques, and even the solution of some topological problems by set theorists.

The general topologist with several courses in mathematical logic behind him is really prepared to take advantage of this movement. A first step in this direction is to become acquainted with a variety of ready-made set theoretic tools which our fellow topologists have been using effectively. The aim of these lectures is to discuss some of these easily grasped tools together with some of the problems they have been used to solve as well as related unsolved problems.

Not all topological problems are set theoretic, of course. We need to recognize to what extent, if any, a problem is set theoretic: to what extent there is a translation into cardinal arithmetic terms. The name of this game is "cardinal functions in topology" which is also the name of a text by István Juhász [22] which I strongly recommend. There are many natural cardinal functions on a topological space.

Let X be a topological space and use the convention that ω stands for finite as well as countable function values. The most obvious function is the *cardinality* of X, called $|X|$; the most useful is the *weight* of X or minimal cardinality of a basis for X, called $w(X)$. A second countable space has weight ω. The *character of X at a point* p, called $\chi(X, p)$, is the minimal cardinality of a basis at p; the *character* of $X = \chi(X) = \sup\{\chi(X, p) | p \in X\}$. First countable spaces have character ω. The *density* of X, called $d(X)$, is the minimal cardinality of a dense subset of X. Separable spaces have density ω. The *Lindelöf number* of X, called $L(X)$, is the minimal cardinal α such that every open cover of X has a subcover of cardinality $\leq \alpha$. Compact and Lindelöf spaces have Lindelöf number ω. Many standard topological properties correspond to the assumption that some natural cardinal function is countable.

Some authors, for example Comfort and Negrepontis [101] and Kunen [95], use $< \alpha$ rather than $\leq \alpha$ in the above definition of Lindelöf number, thus distinguishing between compact and Lindelöf.

Let us give a few other functions which are not quite so obvious but which have recently become standardized for they frequently hide standard pathologies. The *spread* of $X = s(X) = \sup\{|Y| \mid Y \subset X$ and Y is discrete$\}$. The cellularity of $X = c(X) = \sup\{|\mathcal{G}| \mid \mathcal{G}$ is a family of disjoint nonempty open sets$\}$. The *hereditary density of* $X = hd(X) = \sup\{d(Y) | Y \subset X\}$; and the *hereditary Lindelöf number of* $X = hL(X) = \sup\{L(Y) | Y \subset X\}$. The relationship between these four properties is quite delicate and important.

The above functions are defined in terms of sups. Any set theorist would use the smallest cardinal where the property fails rather the sup of those for which it holds. But topologists often prefer the latter definitions because having the countable chain condition correspond to cellularity ω is more natural than ω_1 would be. However, this type of definition immediately raises sup = max problems.

For instance, suppose that the cellularity of a space is α. Is there a family of disjoint nonempty open sets in X of cardinality α? By the useful old (1943) theorem of Erdös and Tarski [23], if X is regular there is such a family unless α is weakly inaccessible: that is, unless α is a regular limit cardinal. It is known to be consistent with the usual axioms for set theory to assume that ω is the only weakly inaccessible cardinal, but it is also felt to be reasonable to assume the existence of uncountable weakly inaccessible cardinals. But regardless of the latter set theoretic problem, the topologist knows exactly when his space can have certain kinds of cellularity difficulties.

This problem also illustrates a fact well-known to logicians and seldom noticed by topologists: ω is a strange cardinal. It is often a special case and not a good indicator for the general theorem. Cardinals come in classes which must frequently be considered separately. Besides ω, regular and singular cardinals must usually be considered separately (and cofinality ω is again often a special case). Beyond this the large cardinals often cause problems just because of their unthinkability. There is a whole theory of large cardinals and the topologist needs to be aware of the possibility for limit cardinals, measureable cardinals, $\cdots$.

Properties like paracompactness and collectionwise normality lend themselves to cardinal function classification. X is *κ-collectionwise normal* provided every closed discrete collection of cardinality κ of closed sets can be separated by disjoint open sets. X is *κ-paracompact* provided every open cover of X of cardinality at most κ has a locally finite open refinement.

The game seems endless, but the point is that it is a useful game. By isolating and comparing these numbers one can use the sophisticated tools available from set theory and combinatorics in a way which is difficult when the numbers are not isolated from the topological structure. We must keep in mind, however, that the number of these functions which will become part of the general mathematical vocabulary is small. We have a responsibility to keep translating these functions and to keep proving theorems which are *topologically* interesting.

We need to be particularly aware of those topological properties which have no cardinal translation. A natural definition for the metricity of a space [81] is the minimal cardinality of a family $\mathfrak{F}$ such that each member of $\mathfrak{F}$ is a discrete set of open sets and $\bigcup\mathfrak{F}$ is a basis. There are a number of cardinal functions which one might call metricity and it is not clear which is most useful.

Normality, which we know from experience is closely tied to set theoretic properties, has no natural translation. In fact none of the separation axioms has such a translation although the cardinal functions are strongly influenced by them. A large part of the complexity in dealing with abstract spaces is related to finding how much a given cardinal function is restricted by a given separation axiom. In the case of normality this is made doubly difficult by the fact that normality is such a second order property that it can often not be decided whether a given topological space is normal or not within the usual axioms for set theory.

Connectedness is another elementary property without a cardinal function translation. However, unlike the separation axioms, connectedness seldom has hidden influence on the more set theoretic properties. The more discrete a set theoretic problem, the more apt one is to be able to solve it. The tendency of the topologist to draw lines rather than discrete sets often blinds him to solutions to set theoretic problems. Here, too, one must be careful not to generalize. The existence of a locally compact normal nonmetrizable Moore space is independent of the usual axioms for set theory. It is consistent that there exists a locally connected normal nonmetrizable Moore space. But Reed and Zenor [24] recently proved that every locally compact locally connected normal Moore space is metrizable. Connectedness is an important topological property which may be relevant even in set theoretic problems. We must keep in mind too that more mathematicians know and care about connectivity than know and care about cellularity. Cellularity is much more important set theoretically.

Our aim in the following chapters will be to discuss a few very specific topological problems from the following areas which seem to be those most effected by set theoretic influence.

(1) Problems which except for separation axioms have pure set theoretic translations:

- (a) cardinal function translations,
- (b) Boolean algebra translations,
- (c) partial order translations.

(2) Metrization problems.

(3) Problems involving the normality and paracompactness of products.

But we will concentrate on the *tools*:

(1) Tree proofs and partition calculus.

(2) Special models for set theory, particularly the contrasting models:

- (a) Martin's axiom together with the negation of the continuum hypothesis,
- (b) $V = L$: Gödel's constructible universe.

II. Ramification Arguments and Partition Calculus

[25] (1) $c \nrightarrow (\omega_1, \omega_1)^2$.

[26] (2) $\omega \rightarrow \omega_n^r$.

[27] (3) $\alpha \rightarrow (\alpha, \omega)^2$.

[28] (4) $(\exp^r \alpha)^+ \rightarrow (\alpha^+)_\alpha^{r+1}$.

[29] (5) $d(X) \leq 2^{s(X)}$.

[29] (6) $|X| \leq 2^{2^{s(X)}}$.

[30] (7) $|X| \leq 2^{(L(X) \circ \chi(X))}$.

[31] (8) If $c(\Pi_{i \in J} X_i) \leq \alpha$ for all finite $J \subset I$, then $c(\Pi_{i \in I} X_i) \leq \alpha$.

The gibberish above is a set of really quite important theorems, the first four from partition calculus and the last four from topology. The topological theorems are proved using partition calculus theorems or techinques which often means using tree type proofs.

In this chapter we first introduce partition calculus notation. This notation is a stone wall barrier which must be crossed in order to make the rich literature of partition calculus available. Almost every problem in partition calculus has been solved; the area is not new; the results are seldom consistency results; the combinatorics of these problems are often present in topological problems; few topologists can even read the theorems.

We use the convention that an ordinal α is the set of all ordinals less than α. A cardinal α is the smallest ordinal of cardinality α. We use $\mathrm{cf}(\alpha)$ to stand for the cofinality of α. And when we speak of α as a topological space we give it the order topology.

If S is a set and r is a positive integer then $[S]^r$ is the standard partition calculus notation for $\{X \subset S \mid |X| = r\}$. A *partition* of $[S]^r$ is a family $\{F_i\}_{i \in I}$ whose union is $[S]^r$. A set $A \subset S$ is said to be *homogeneous* (for this partition) provided there is an $i \in I$ with $[A]^r \subset F_i$. Let α, β, γ, and r be cardinals with r a positive integer, α infinite, $\gamma < \alpha$, and $r < \beta \leq \alpha$. Then, in partition calculus language, $\alpha \rightarrow (\beta)_\gamma^r$ is the statement:

Given a set S with $|S| = \alpha$ and a partition $\{F_i\}_{i \in I}$ of $[S]^r$ where $|I| = \gamma$, then there is an $A \subset S$ and $i \in I$ such that $|A| = \beta$ and $[A]^r \subset F_i$.

$\alpha \rightarrow (\beta_1, \cdots, \beta_j)^r$ says that if S is a set with $|S| = \alpha$ and $(F_1, \cdots, F_j)$ is a partition of $[S]^r$, then there is an $A \subset S$ and an $i \leq j$ with $|A| = \beta_i$ and $[A]^r \subset F_i$.

The notation is awful; but the concepts are useful. It is necessary to have seen the notation and thought about the concepts to gain access to these tools. See Juhász [22] for a pretty treatment.

(1) *Let us prove* SIERPINSKI'S THEOREM: $c \not\to (\omega_1, \omega_1)^2$.

Let $<^*$ be a well ordering of the real numbers. Let $F_1 = \{(a, b) \in \text{reals} \mid a < b$ and $a <^* b\}$ and let $F_2 = ([\text{reals}]^2 - F_1)$. Neither F_1 nor F_2 can contain an uncountable homogeneous set for the reals contain no uncountable well ordered sequence.

(2) *Let us prove* RAMSEY'S THEOREM: $\omega \to (\omega)^r_n$.

Assume an infinite set S and a partition $\{F_i\}_{i<n}$ of the set $[S]^r$ of all r-tuples in S. We prove the theorem by induction on r. If $r = 1$ there is clearly an F_i which is infinite. Assume the theorem for $r = k$ and suppose $r = k + 1$. Choose $s_0 \in S$. Since the theorem is true for $r = k$, there is an infinite $S_0 \subset S$ and an $i_0 < n$ such that, if $X \in [S_0]^k$, $(\{s_0\} \cup X) \in F_{i_0}$. Choose $s_1 \in S_0$. By induction we can clearly choose $S_0 \supset S_1 \supset \cdots$ and for each $j \in \omega_0$ an $s_j \in S_{j-1}$ and an $i_j < n$ such that, for all $X \in [S_j]^k$, $(\{s_j\} \cup X) \in F_{i_j}$. Since n is finite there is an $i < n$ and an infinite subset J of ω such that $i_j = i$ for all $j \in J$. Clearly $\{s_j\}_{j \in J}$ is an infinite subset of S all of whose $(k + 1)$-tuples belong to F_i.

This 1930's proof is beautifully linear.

(3) *Let us prove* ERDOS' THEOREM: $\alpha \to (\alpha, \omega)^2$.

Assume a set S of cardinality α and a partitioning of the unordered pairs from S into two classes F_1 and F_2. We assume that there is no infinite homogeneous set for F_2.

If $T \subset S$ and $s \in S$, let $F_2(T, s) = \{t \in T \mid (s, t) \in F_2\}$.

First observe that there is a $T \subset S$ with $|T| = \alpha$ such that, for all $t \in T$, $|F_2(T, t)| < \alpha$. Otherwise, we can choose by induction, for each $n \in \omega$, $T_n \subset S$ and $t_n \in T_n$ such that $|T_n| = \alpha$ and $T_{n+1} = F_2(T_n, t_n)$. But then $\{t_n\}_{n \in \omega}$ would be an infinite set all of whose pairs belong to F_2.

We next observe that $\alpha \to (\alpha, \omega)^2$ for α regular. Let M be a maximal subset of T with $[M]^2 \subset F_1$. Since $|T| = \alpha$ and, for all $t \in M$, $|F_2(T, t)| < \alpha$, if α is regular, $|M| = \alpha$.

So assume that α is singular and that for regular β, $\beta \to (\beta, \omega)^2$. There is a family Λ of $\mathrm{cf}(\alpha)$ regular cardinals greater than $\mathrm{cf}(\alpha)$ whose union is α. Without loss of generality we assume that $T = S = \alpha$.

By transfinite induction, for each $\beta < \mathrm{cf}(\alpha)$, choose $\lambda_\beta \in \Lambda$ and $S_\beta \subset \lambda_\beta$ as follows. Suppose λ_γ and S_γ have been chosen for all $\gamma < \beta$. Choose $\delta \in \Lambda$ such that $\delta > \sup\{\lambda_\gamma\}_{\gamma<\beta}$; let $D = |\delta - \bigcup_{\gamma<\beta}\lambda_\gamma|$ and observe that, since δ is regular, $|D| = \delta$. There is a $\lambda_\beta \in \Lambda$ and $E \subset D$ such that $|E| = \delta$ and $(s, t) \in F_1$ for all $s \in E$ and $t \geq \lambda_\beta$. Since δ is regular $\delta \to (\delta, \omega)^2$ and we can choose $S_\beta \subset E$ with $|S_\beta| = \delta$ and $[S_\beta]^2 \subset F_1$.

If $A = \bigcup_{\beta<\lambda} S_\beta$, $|A| = \alpha$ and $[A]^2 \subset F_1$. Thus $\alpha \to (\alpha, \omega)^2$.

This 1940's proof is nasty but still linear.

(4) *We prove the* ERDÖS-RADO THEOREM FOR $r = 1$: $(2^{\alpha})^+ \to (\alpha^+)^2_\alpha$.

Suppose that we have a set S with $|S| > 2^\alpha$ and that $\{F_\beta\}_{\beta<\alpha}$ is a partitioning of the unordered pairs from S. We use induction to build a tree. Arbitrarily choose $p \in S$.

For each $\gamma < \alpha^+$ and function $f: \gamma \to \alpha$, we define $S_f \subset S$ and $p_f \in S$ by induction as follows.

Suppose that $\gamma < \alpha^+$ and that S_g and p_g have been defined for all $g: \delta \to \alpha$ where $\delta < \gamma$ and that $p_h \notin S_g$ for $h = g \upharpoonright \eta$ with $\eta < \delta$.

If $\gamma = 0$ then $f = \emptyset$. Define $S_\emptyset = S$.

If $\gamma = \delta + 1$, define $S_f = \{x \in S_{f\upharpoonright\delta} | (x, p_{f\upharpoonright\delta}) \in F_{f(\delta)}\}$.

If γ is a limit ordinal, define $S_f = \bigcap_{\delta<\gamma} S_{f\upharpoonright\delta}$.

If $S_f \neq \emptyset$ choose $p_f \in S_f$; otherwise define $p_f = p$.

Let $F = \{\text{function } f \mid \text{for some } \gamma < \alpha^+, f \text{ maps } \gamma \text{ into } \alpha\}$; $|F| \leq 2^\alpha$. So there is a $y \in S$ such that $y \neq p_f$ for any $f \in F$. By induction on γ, for each $\gamma < \alpha^+$, choose $f^\gamma: \gamma \to \alpha$ such that $y \in S_{f^\gamma}$ and $\delta < \gamma$ implies that $f^\gamma \upharpoonright \delta = f^\delta$.

For $\beta < \alpha$, define $G_\beta = \{\gamma < \alpha^+ | f^{\gamma+1}(\gamma) = \beta\}$. For some $\beta < \alpha$, $|G_\beta| = \alpha^+$. If $A = \{p_{f^\gamma}\}_{\gamma \in G_\beta}$, $|A| = \alpha^+$ and $[A]^2 \subset F_\beta$. Thus our proof that $(2^\alpha)^+ \to (\alpha^+)^2_\alpha$ is complete.

Observe that there is a natural partial order on the members of F. Define $f \leq g$ if the domain γ of f is contained in the domain of g and $g \upharpoonright \gamma = f$. Since the predecessors of any $g \in F$ are well ordered, $(F, \leq)$ is a tree. The proof depended on showing that the tree had a chain which meets every level of the tree (called having the *tree property*). This type of argument is called a *ramification* argument and is often useful.

Let us prove the ERDÖS-RADO THEOREM: $(\exp^r \alpha)^+ \to (\alpha^+)^{r+1}_\alpha$.

We prove the theorem by a ramification argument and induction on r. Assume that $r > 1$, $\lambda = \exp^{r-1}\alpha$, and $\lambda^+ \to (\alpha^+)^r_\alpha$. We prove that $(2^\lambda)^+ \to (\alpha^+)^{r+1}_\alpha$.

Suppose that S is a set with $|S| > 2^\lambda$ and that $\{F_\beta\}_{\beta<\alpha}$ is a partitioning of the unordered $(r+1)$-tuples from S. Arbitrarily choose $(p_0, \cdots, p_{r-1}) \subset S$.

For each $\gamma < \lambda^+$ and function $f: [\gamma]^r \to \alpha$, we define $S_f \subset S$ and $p_f \in S$ by induction as follows.

Suppose that $\gamma < \lambda^+$ and that S_g and p_g have been defined for all $g: [\delta]^r \to \alpha$ where $\delta < \gamma$ and that $p_h \notin S_g$ for any $h = g \upharpoonright [\eta]^r$ with $\eta < \delta$. (If $\delta < \gamma < r$ we think of $f: [\delta]^r \to \alpha$ as different from $g: [\gamma]^r \to \alpha$.)

If $\gamma < r$, define $S_f = S - (p_0, \cdots, p_{r-1})$.

If $r \leq \gamma = \delta + 1$, define $S_f = \{x \in S_{f\upharpoonright[\delta]^r} - \{p_{f\upharpoonright[\delta]^r}\} | \forall R \in [\gamma]^r$, the $(r+1)$-tuple $(\{x\} \cup \{p_{f\upharpoonright[\eta]^r} | \eta \in R\}) \in F_{f(R)}\}$.

If γ is a limit ordinal, define $S_f = \bigcap_{\delta<\gamma} S_{f\upharpoonright[\delta]^r}$.

If $\gamma < r$, define $p_f = p_\gamma$. If $\delta \geq r$ and $S_f \neq \emptyset$, choose $p_f \in S_f$. Otherwise define $p_f = p_0$.

Let $F = \{\text{function } f \mid \text{for some } \gamma < \lambda^+, f \text{ maps } [\gamma]^r \text{ into } \alpha\}$; $|F| \leq \lambda^+ \cdot \alpha^\lambda = 2^\lambda$. So there is a $y \in S$ such that $y \neq p_f$ for any $f \in F$. By induction on γ, for each $\gamma < \lambda^+$, choose $f^\gamma: [\gamma]^r \to \alpha$ such that $y \in S_{f^\gamma}$ and $\delta < \gamma$ implies that $f^\gamma \upharpoonright [\delta]^r = f^\delta$.

Let $T = \{p_{f^\gamma} \mid \gamma < \lambda^+\}$; $|T| = \lambda^+$. Let $\{I_\beta\}_{\beta<\alpha}$ be the partitioning of the r-tuples in T defined by $R \in I_\beta$ provided $f^\gamma(R) = \beta$ for some $\gamma < \lambda^+$. Since $\lambda^+ \to (\alpha^+)^r_\alpha$, there is $T' \subset T$ with $|T'| = \alpha^+$ and $[T']^r \subset I_\beta$. But then $[T']^{r+1} \subset F_\beta$.

We now turn to the topological theorems.

(5) *We prove the* JUHÁSZ-HAJNAL THEOREM: $d(X) \leq 2^{s(X)}$.

Let $s(X) = \alpha$; we make use of the Erdös theorem $\alpha^+ \to (\alpha^+, \omega)^2$.

Assume that $d(X) > 2^\alpha$. By simple transfinite induction we choose a sequence $P = \{p_\beta\}_{\beta<(2^\alpha)^+}$ such that $p_\beta \in X - \overline{\{p_\gamma\}_{\gamma<\beta}}$.

We want to find a sequence $\{q_\beta\}_{\beta<\alpha^+} \subset P$ such that $q_\beta \notin \overline{\{q_\gamma\}_{\gamma>\beta}}$. To do this we use a ramification argument. For each $\gamma < \alpha^+$, let $F_\gamma = \{f : \gamma \to (0, 1)\}$ and let $F = \bigcup_{\gamma<\alpha^+} F_\gamma$.

For each $\gamma < \alpha^+$ and function $f \in F_\gamma$, we define a closed set $P_f \subset P$ as follows.

Suppose that $\gamma < \alpha^+$ and that P_g has been defined for all $g \in F_\delta$ with $\delta < \gamma$.

If $\gamma = 0$, then $F_\gamma = \emptyset$. Define $P_\emptyset = P$.

If $\gamma = \delta + 1$ and f and g are members of F_γ such that $f \upharpoonright \delta = g \upharpoonright \delta$ and $f(\gamma) \neq g(\gamma)$, then

(a) if $|P_{f\upharpoonright\delta}| \leq 1$, define $P_f = P_g = \emptyset$, but

(b) if $|P_{f\upharpoonright\delta}| > 1$, let P_f and P_g be proper closed subsets of $P_{f\upharpoonright\delta}$ whose union is $P_{f\upharpoonright\delta}$.

Since $|F| = 2^\alpha$ and $|P| > 2^\alpha$, there is a $p \in P$ such that $\{p\} \neq P_f$ for any f. For each $\gamma < \alpha^+$, choose $f^\gamma \in F_\gamma$ such that $p \in P_{f^\gamma}$ and $f^\gamma \upharpoonright \delta = f^\delta$ for all $\delta < \gamma$. For each $\gamma < \alpha^+$, choose $q_\gamma \in P_{f^\gamma} - P_{f^{\gamma+1}}$. Clearly $q_\gamma \notin \overline{\{q_\delta\}_{\delta>\gamma}}$.

Let $Q = \{q_\gamma\}_{\gamma<\alpha^+}$. Partition $[Q]^2$ into two classes F_1 and F_2 as follows. If $(q_\delta, q_\gamma) \in [Q]^2$ and $\delta < \gamma$ and $q_\delta = p_\rho$ and $q_\gamma = p_\eta$, then $(q_\delta, q_\gamma) \in F_1$ if $\rho < \eta$ and $(q_\delta, q_\gamma) \in F_2$ if $\rho > \eta$. There can be no infinite set $A \subset Q$ with $[A]^2 \subset F_2$ for there is no infinite decreasing set of ordinals. So since $\alpha^+ \to (\alpha^+, \omega)^2$, there is a set $A \subset Q$ with $|A| = \alpha^+$ and $[A]^2 \subset F_1$. Since A is discrete this contradicts $s(X) = \alpha$.

(6) *Let us prove* JUHÁSZ AND HAJNAL'S THEOREM: $|X| \leq 2^{2^{s(X)}}$.

We make use of the Erdös-Rado theorem $2^{2^\alpha} \to (\alpha^+)^3_4$.

Well order X by $<$ and, for each x and y in X, find disjoint open sets U_{xy} and V_{xy} with $x \in U_{xy}$ and $y \in V_{xy}$. Partition $[X]^3$ into four classes F_1, F_2, F_3, and F_4: If $x < y < z$ put

$$(x, y, z) \in F_1 \quad \text{if} \quad z \in U_{xy} \text{ and } x \in V_{yz},$$
$$(x, y, z) \in F_2 \quad \text{if} \quad z \notin U_{xy} \text{ and } x \in V_{yz},$$
$$(x, y, z) \in F_3 \quad \text{if} \quad z \notin U_{xy} \text{ and } x \notin V_{yz},$$
$$(x, y, z) \in F_4 \quad \text{if} \quad z \in U_{xy} \text{ and } x \notin V_{yz}.$$

Let $\alpha = s(X)$ and suppose that $|X| > 2^{2^\alpha}$. By the Erdös-Rado theorem there is a subset of Y of X with $|Y| = \alpha^+$ and $[Y]^3$ is a subset of one F_1, F_2, F_3, and F_4. Let $Z = \{t \in Y \mid t$ has an immediate predecessor in Y under $<\}$. We claim that Z is discrete. Since $|Z| = \alpha^+$ this contradicts $\alpha = s(X)$. We prove that Z is discrete if $[Y]^3 \subset F_2$; the other cases are similar.

If $t \in Z$ then let z be the immediate successor of t in Z. Clearly t is not in the closure of $\{s \in Z \mid s < t\}$, for $s < t$ implies $s \in V_{tz}$; hence $s \notin U_{tz}$ while $t \in U_{tz}$. Also t is not in the closure of $\{s \in Z \mid t < s\}$, for $t < z < s$ implies $s \notin U_{tz}$ while $t \in U_{tz}$.

(7) *Let us prove* ARHANGELSKII'S THEOREM: $|X| \leq 2^{\chi(X) \circ L(X)}$.

This theorem answered the 50 year old question of Alexandroff as to whether every first countable compact space has cardinality $\leq c$.

A *free sequence* in a space X is a subset $\{x_\alpha\}_{\alpha<\gamma}$ of X such that, for all $\delta < \gamma$, $\{x_\beta\}_{\beta<\delta}$ and $\{x_\beta\}_{\beta\geq\delta}$ have disjoint closures. Let $\lambda = \chi(X) \circ L(X)$.

LEMMA 1. *If $Y \subset X$ and $|Y| \leq \lambda$, then $|\overline{Y}| \leq 2^\lambda$.*

LEMMA 2. *If Z is a closed subset of X and $|Z| \leq 2^\lambda$, then $X - Z$ is the union of 2^λ closed sets.*

LEMMA 3. *There is no free sequence in X of cardinality $> \lambda$.*

For $x \in X$, let $\{U_\alpha(x)\}_{\alpha<\lambda}$ be a local basis at x; this is possible since $\chi(X) \leq \lambda$.

PROOF OF LEMMA 1. For $y \in \overline{Y}$, let $V_\alpha(y) = U_\alpha(y) \cap Y$ and let $\mathcal{O}(y) = \{V_\alpha(y)\}_{\alpha<\lambda}$. If $z \in \overline{Y}$ and $z \neq y$, then $\mathcal{O}(z) \neq \mathcal{O}(y)$. Since the number of subsets of Y is at most 2^λ, $|\overline{Y}| \leq 2^{\lambda \circ \lambda} = 2^\lambda$.

PROOF OF LEMMA 2. Let $\mathcal{G} = \{U_\alpha(z) \mid z \in Z$ and $\alpha < \lambda\}$. Then $|\mathcal{G}| \leq \lambda \circ 2^\lambda = 2^\lambda$. Let $\mathcal{G}^* = \{$unions of $\leq L(X)$ members of $\mathcal{G}\}$. Then $|\mathcal{G}^*| \leq 2^\lambda$. Let $Q = \{X - G \mid G \in \mathcal{G}^*$ and G covers $Z\}$. Clearly Q is a family of closed subsets of $X - Z$ and $|Q| \leq 2^\lambda$. Since Q covers $X - Z$ the lemma is proved.

PROOF OF LEMMA 3. Suppose that $S = \{x_\alpha\}_{\alpha<\lambda^+}$ is a free sequence in X. For each $\alpha < \lambda^+$, let $U_\alpha = X - \overline{\{x_\beta\}_{\beta\geq\alpha}}$. If $x \notin \overline{S}$, then $x \in U_0$. If $x \in \overline{S}$ choose $s_{x\gamma} \in S \cap U_\gamma(x)$ for each $\gamma < \lambda$. Since $x \in \overline{\{s_{x\alpha}\}_{\alpha<\lambda}}$ and S is free, $x \in U_\alpha$ for some $\alpha < \lambda^+$. Therefore $\{U_\alpha\}_{\alpha<\lambda^+}$ is an open cover of X. Since $L(X) \leq \lambda$, this is a contradiction.

PROOF OF ARHANGELSKII'S THEOREM FROM THE LEMMAS. Assume that $|X| > 2^\lambda$. We use a ramification argument to choose a free sequence of cardinality λ^+ in X. Define $F_\gamma = \{f\colon \gamma \to 2^\lambda\}$ for each $\gamma < \lambda^+$. Choose $x \in X$.

For each $\gamma < \lambda^+$ and $f \in F_\gamma$, we define X_f and p_f by induction on γ as follows.

Suppose that $\gamma < \lambda^+$ and that X_g and p_g have been defined for all $g \in F_\delta$ for $\delta < \gamma$ and that $p_h \notin X_g$ for any $h \in F_\eta$ with $\eta < \delta$.

If $\gamma = 0$, $F_\gamma = \{\emptyset\}$. Define $X_\emptyset = X$.

If $\gamma = \beta + 1$ and $g \in F_\beta$, let $F_g = \{f \in F_\gamma \mid f \restriction \beta = g\}$. Let $Y_g = \{p_h \mid h = g \restriction \eta$ for some $\eta \leq \beta\}$. Since $|Y_g| \leq \lambda$, by Lemma 1, $|\overline{Y}_g| \leq 2^\lambda$. By Lemma 2, there is a family $\{H_\alpha\}_{\alpha<2^\lambda}$ of closed subsets of X whose union is $X - \overline{Y}_g$. For $f \in F_g$, define $X_f = H_{f(\beta)}$.

If γ is a limit ordinal and $f \in F_\gamma$, define $X_f = \bigcap\{X_g \mid f \restriction \beta = g$ for some $\beta < \gamma\}$.

If $f \in F_\gamma$ and $X_f \neq \emptyset$, choose $p_f \in X_f$. If $X_f = \emptyset$, define $p_f = x$.

Since $|\bigcup_{\gamma<\lambda^+} F_\gamma| = 2^\lambda$ and $|X| > 2^\lambda$, there is a $y \in X$ such that $y \neq p_f$ for any

$f \in \bigcup_{\gamma<\lambda^+} F_\gamma$. For each $\gamma < \lambda^+$, choose $f^\gamma \in F_\gamma$ by induction on γ such that $\gamma \in X_{f^\gamma}$ and $f^\gamma \restriction \delta = f^\delta$ for all $\delta < \gamma$. Then $\{p_{f^\gamma}\}_{\gamma<\lambda^+}$ is a free sequence contradicting Lemma 3.

(8) Our last theorem is in quite a different style but it gives us an opportunity to make two observations.

(a) Cardinal functions of products are often determined by the same functions of subproducts of some size.

(b) To prove a theorem of the type described in (a) what you usually need is a lemma about a "delta system."

A "delta system" is a partition calculus concept quite different from the ones we discussed earlier but even more broadly applicable. If α, β, and λ are cardinals, $\delta \to \Delta(\alpha, \lambda)$ is the statement: given a family F of δ sets each of cardinality $< \alpha$, there is a delta system G in F of cardinality λ; that is $G \subset F$, $|G| = \lambda$, and there is a set J such that every pair of members of G intersects in J. An easy, frequently used theorem is that, if δ is uncountable and regular, then $\delta \to \Delta(\omega, \delta)$.

PROOF OF NOBLE AND ULMER'S THEOREM. If $c(\Pi_{i\in J} X_i) \le \alpha$ for all finite $J \subset I$, then $c(\Pi_{i\in I} X_i) \le \alpha$.

Suppose that α is infinite and $\{V_\beta\}_{\beta<\alpha^+}$ is a family of disjoint nonempty open sets in $\Pi_{i\in I} X_i$. For each $\beta < \alpha^+$ there is a finite subset F_β of I and a nonempty basic open set $\Pi_{i\in I} W(\beta, i) \subset V_\beta$ such that $W(\beta, i) \ne X_i$ only if $i \in F_\beta$. Since α^+ is an uncountable regular cardinal, there is $G \subset \{F_\beta\}_{\beta<\alpha^+}$ with $|G| = \alpha^+$ and $J \subset I$ such that $F_\beta \cap F_\gamma = J$ for all F_β and $F_\gamma \in G$. Let $B = \{\beta < \alpha^+ \mid F_\beta \in G\}$. If $J \ne \emptyset$, then $\{\Pi_{i\in J} W(\beta, i)\}_{\beta\in B}$ is a family of α^+ nonempty open sets in $\Pi_{i\in J} X_i$. Since $c(\Pi_{i\in J} X_i) \le \alpha$, there must be β and γ in B and $\{x_i\}_{i\in J} \in (\Pi_{i\in J} W(\beta, i) \cap \Pi_{i\in J} W(\gamma, i))$. If $J = \emptyset$ choose β and γ in B arbitrarily. If $i \in (F_\beta - J)$ choose $x_i \in W(\beta, i)$; if $i \in (F_\gamma - J)$ choose $x_i \in W(\gamma, i)$; if $i \in (I - (F_\beta \cup F_\gamma))$ choose $x_i \in X_i$. Then $\{x_i\}_{i\in I} \in (\Pi_{i\in I} W(\beta, i) \cap \Pi_{i\in I} W(\gamma, i))$ which contradicts the fact that these sets were disjoint.

III. Souslin Trees and Martin's Axiom

The existence of a Souslin line implies the following:

[44] (1) There is a compact, connected, linear order topology, hereditarily Lindelöf and ccc, first countable, perfectly normal, nonseparable space.

[45] (2) There is a nonspecial Aronszajn tree.

[35] (3) There is a ccc space whose square is not ccc.

[34], [48] (4) There is a (first countable) hereditarily separable, real compact, cardinality ω_1 Dowker space.

(MA + $\neg$CH) implies the following:

[50] (5) There is no Souslin line.

[37] (6) Perfectly normal compact spaces are hereditarily separable.

(K. Kunen [22]) (7) Every product of ccc spaces is ccc.

[39] (8) Every Aronszajn tree is special.

So far we have tried to make two points. The first is that abstract space topology is remarkably set theoretic in nature. The second is that partition calculus and other purely combinatorial cardinal arithmetic techniques are still used to prove interesting theorems. We now wish to make a third point which is that some simple important apparently purely topological problems cannot be decided using the usual axioms for set theory. It is very important to know when the problem we are working on is of this class.

By the "usual axioms for set theory" we mean the nine simple axioms of Zermelo-Fraenkel set theory, axioms so basic as to seem impossible to deny, together with the axiom of choice. The axiom of choice is a major assumption but one which both logicians and topologists are usually willing to make. These are the rules under which mathematicians usually work. A proof is an argument which uses only these rules. Using forcing one can build many highly specialized models for set theory.

We concentrate on two models for set theory which topologists have found particularly useful.

In set theory language, V is the class of all sets; L is the class of constructible sets. Gödel's axiom of constructibility says that $V = L$.

Jensen proved that, if $V = L$, then there is a Souslin line. Later it was seen that the proof could be broken into two parts: $V = L$ implies $\diamondsuit$ and $\diamondsuit$ implies there is a Souslin line (as well as CH). $\diamondsuit$ is a combinatorial statement that has proved especially useful and easy to apply.

Martin's axiom together with not-CH, usually called (MA + $\neg$CH), is equally versatile and implies there is *no* Souslin line.

Our aim for this chapter is to discuss trees in general, particularly Cantor trees, Aronszajn trees, and Souslin trees. We will also introduce Martin's axiom and show some of the consequences of (MA + $\neg$CH) on the nature of these trees. Trees are a very fundamental set theoretic concept partly because models for set theory are built as trees.

A *tree* [102] is a partially ordered set $(T, \leq)$ such that, for each $t \in T$, the set of all predecessors of t in T is well ordered by $\leq$. If α is an ordinal, the αth *level* of T is the set of all $t \in T$ such that the set of all predecessors of t under $\leq$ is order isomorphic to α under its natural order. The *height* of T is the smallest α such that αth level of T is empty. A *chain* in T is a totally ordered subset of T (in a tree chains are well ordered). An *antichain* in T is a pairwise unordered subset of T. We say that T *has the tree property* if there is a chain which meets every level of T. An open interval in T is a subset J of T such that, for some $t \in T$, either $J = \{t\}$ and t belongs to the first level of T, or there is an $s < t$ in T such that $J = \{x \in T \mid s < x \leq t\}$.

The *tree topology* on a tree $(T, \leq)$ is obtained by using the set of all open intervals in T as a basis for a topology on T; the tree topology is always Hausdorff and regular. On a partially ordered set $(P, \leq)$ there is a different topology which is called the *partial order topology* and is obtained by using $\{\{x \in P \mid x \geq p\} \mid p \in P\}$ as a basis. This topology is seldom Hausdorff but is the topology used by logicians to build models for set theory.

We construct the *Cantor tree* C by induction. The first level of C is the interval $[0, 1]$; the second level is the pair of intervals $[0, 1/3]$ and $[2/3, 1]$; $\cdots$. The ωth level of C is the Cantor set C' on which these intervals close down and the partial order is by inclusion. If $\kappa \leq c$ is a cardinal and B is a subtree of C containing $C - C'$ with $|B| = \kappa$, we call B a *κ-Cantor tree.* A κ-Cantor tree B, for uncountable κ, is a separable, locally compact, nonmetrizable, Moore space.

A ω_1-Cantor tree B is normal if and only if each pair of complementary subsets of $B \cap C'$ can be separated by disjoint open sets, If we assume $c = 2^{\omega} < 2^{\omega_1}$, which follows from the continuum hypothesis, then B is not normal. But (MA + $\neg$CH) implies that B is normal, a fundamental set theoretic fact which we shall presently prove.

A tree $(T, \leq)$ is called an *Aronszajn tree* provided T is uncountable but every chain and level of T is countable. An Aronszajn tree is called *special* provided it is the union of countably many antichains. Aronszajn [41] and Jones [42] independently constructed special Aronszajn trees. These trees can be used to construct compact, connected, linearly ordered spaces which are the union of ω_1 nowhere dense sets. Details related to this paragraph and the next can be found in [43].

Suppose that S is a connected, linear order topology space without a first or last point. If S is separable, S is a line. Countable cellularity is standardly abbreviated ccc which really stands for countable chain condition. If S has ccc but is not separable, S is called a *Souslin line*. A *Souslin tree* is an uncountable tree in which every chain and antichain is countable. The existence of a Souslin line is equivalent [45] to the existence of a Souslin tree. One of the earliest consistency results in topology was that the existence of a Souslin line is independent of the usual axioms for set theory [38], [50].

Martin's axiom is not something which excites D. A. Martin very much. He observed that various constructions of models for set theory in which there were no Souslin lines had something in common; and this common denominator became known as Martin's axiom. It can be stated in many equivalent forms; the partial order form is the most useful, but topologists find the Baire category form the most natural.

One version of the Baire category theorem states that no compact Hausdorff space is the union of countably many nowhere dense sets.

Martin's axiom states that [36] *no compact ccc Hausdorff space is the union of less than c nowhere dense sets.*

Under CH, Martin's axiom is the Baire category theorem with an unnecessary ccc hypothesis. But, recall that an Aronszajn tree yields a compact Hausdorff space which is the union of ω_1 nowhere dense sets. So if CH is not true, the ccc in the hypothesis is necessary. The assumption which (is consistent with the usual axioms for set theory and) is most interesting is (MA + $\neg$CH), although a remarkable number of the consequences of CH are consequences of the weaker Martin's axiom.

(1) A Souslin line S contains at most countably many maximal separable nontrivial closed subintervals. Collapse each of these intervals to a point and add a first and last point to the space. The resulting space S' is compact, connected, hereditarily Lindelöf, hereditarily ccc, first countable, perfectly normal, and not separable. To see that S' is the union of ω_1 nowhere dense subsets we use the standard construction of a Souslin tree from a Souslin line.

By transfinite induction we choose for each $\alpha < \omega_1$, a countable subset A_α of S' which has nonempty intersection with every component of $S' - \bigcup_{\beta<\alpha} A_\beta$. Since S' has no separable open intervals, each A_α is nowhere dense, and since S' has ccc, $\bigcup_{\alpha<\omega_1} A_\alpha = S'$. Thus S' is the union of ω_1 nowhere dense subsets.

(2) In the construction above, let T_α be the set of all components of $S' - \overline{\bigcup_{\beta<\alpha} A_\beta}$ and let $T = \bigcup_{\alpha<\omega_1} T_\alpha$. Partially order T by reverse inclusion. Then $(T, \leq)$ is a Souslin tree and thus a nonspecial Aronszajn tree.

(3) Observe that if $U \in T_\alpha$, above, then there are disjoint U_1 and U_2 in $T_{\alpha+1}$ with $U_1 \subset U$ and $U_2 \subset U$. Let $G = \{U_1 \times U_2 \mid U \in T\}$. Then $|G| = \omega_1$ and G is a family of disjoint open sets in $S' \times S'$. Thus, although S' has ccc, $S' \times S'$ does not.

(4) A Dowker space is a normal space which is not countably paracompact. The only known example of a Dowker space [47] is not real compact, has cardinality and cellularity $(\omega_\omega)^\omega$ and character ω_ω. However a Souslin tree can be used to construct a real compact Dowker space of cardinality ω_1 which is hereditarily separable (or first countable) [48].

If κ is a cardinal, we define a tree $(T, \leq)$ to be a *κ-Souslin tree* provided $|T| = \kappa$ but T has no chain or antichain of cardinality κ. A Souslin tree is thus an ω_1-Souslin tree. There is no ω-Souslin tree. For singular κ there are trivally κ-Souslin trees. For other cardinals κ it is consistent with the usual axioms for set theory that there exist (or not exist) κ-Souslin trees.

The Dowker space described in [47] (and Chapter IX (2)) is constructed using a trivial ω_ω Souslin tree. Any κ-Souslin tree where κ is not a singular cardinal or the successor of a singular cardinal can be used to construct a Dowker space of cardinality κ [49] whose cardinal functions are bounded by κ. The question of whether it is possible, without assuming the existence of a Souslin tree, to construct a Dowker space any of whose cardinal functions are small is now solved assuming CH.

(5) In (1) we proved that a Souslin line can be used to construct a compact ccc space which is the union of ω_1 nowhere dense sets; therefore, (MA + $\neg$CH) implies that there is no Souslin line.

(6) Ponomarev raised the question of the existence of other perfectly normal non-separable, compact spaces and we now give Juhász's answer.

(a) *If X is a perfectly normal compact space, then X is first countable and ccc.* To see the latter, assume that $\{U_\alpha\}_{\alpha<\omega_1}$ is a family of disjoint nonempty open sets. For each $\alpha < \omega_1$ choose $p_\alpha \in U_\alpha$. Let $H = \overline{\bigcup_{\alpha\in\omega_1} U_\alpha} - \bigcup_{\alpha\in\omega_1} U_\alpha$. Then $H = \bigcap_{n\in\omega} W_n$ where each W_n is open. Thus there is an $n \in \omega$ and an infinite subset M of ω_1 such that $\{p_\alpha | \alpha \in M\} \subset X - W_n$. $\{p_\alpha | \alpha \in M\}$ is an infinite set in the compact space X without a limit point which is impossible.

(b) *If X is a compact ccc space, λ is a regular cardinal, and $\omega < \lambda < c$, then Martin's axiom implies that any family $G = \{U_\alpha\}_{\alpha<\lambda}$ of open sets has a cardinality λ subfamily with nonempty intersection.*

Let F be a maximal family of disjoint nonempty open subsets of X each of which meets less than λ members of G. Since X has ccc, F is countable. So there is a member V of G which does not meet any member of F. Clearly, if W is any open subset of $\overline{V}$, W intersects λ members of G. For $\beta < \lambda$, define $H_\beta = \overline{V} - \overline{\bigcup_{\beta<\alpha} U_\alpha}$. Then H_β is nowhere dense in the compact ccc space $\overline{V}$. Thus there is a $v \in \overline{V}$ such that $v \notin \bigcup_{\beta<\lambda} H_\beta$ and hence v belongs to λ U_α's.

(c) (MA + $\neg$CH) *implies that every compact, perfectly normal space Y is hereditarily separable.*

If Y is not hereditarily separable we can choose a subset $\{x_\alpha\}_{\alpha<\omega_1}$ of Y by induction such that $x_\alpha \notin \overline{\{x_\beta\}_{\beta<\alpha}}$. Let $X = \overline{\{x_\alpha\}_{\alpha<\omega_1}}$. By (a), X is a compact, ccc, first countable space. For $\beta < \omega_1$, define $U_\beta = X - \overline{\{x_\alpha\}_{\alpha<\beta}}$. By (b), there is a $v \in \bigcap_{\beta<\omega_1} U_\beta$ but this is impossible since X is first countable.

(7) Before proving (7) let us discuss the fundamental concept of a *Stone space* [51].

Suppose that (X, T) is a topological space which need not even be Hausdorff! If $U \in T$, let U^* be the interior of $\overline{U}$; U^* is called a *regular* open set. Observe that, if

$U \cap V = \emptyset$, then $U^* \cap V^* = \emptyset$. Let $S(X, T)$ be the set of all ultrafilters on $\{U^* \mid U \in T\}$. If $U \in T$, let $U' = \{A \in S(X, T) \mid U^* \in A\}$. Note that $U \in T$ and $A \in S(X, T)$ imply that either $U^* \in A$ or $X - \overline{U}^* \in A$. The *Stone space* of (X, T) is $S(X, T)$ topologized by using $\{U' \mid U \in T\}$ as a basis. The Stone space of a space (X, T) is compact, extremally disconnected, and Hausdorff, and, if (X, T) has ccc, then the Stone space of X has ccc.

Let us prove that (MA + $\neg$CH) *implies that* if $\{X_\beta\}_{\beta \in B}$ *is a family of ccc spaces, then* $X = \Pi_{\beta \in B} X_\beta$ *has ccc.*

Suppose that $\{U_\alpha\}_{\alpha < \omega_1}$ is a family of disjoint, nonempty, basic open sets in X. For each α there is a finite $F_\alpha \subset B$ such that $U_\alpha = \Pi_{\beta \in B} W_{\alpha\beta}$ and $W_{\alpha\beta} \neq X_\beta$ only if $\beta \in F_\alpha$. By a delta system argument choose an uncountable subset D of ω_1 and a finite subset J of B such that $F_\beta \cap F_\gamma = J$ for all β and γ in D.

For each $\beta \in J$, let S_β be the Stone space of X_β and define U' as above. Let $J = \beta_1, \cdots, \beta_j$. Since S_β is compact, Hausdorff and ccc, we can choose (using (6)(b)) uncountable subsets $D_1 \supset \cdots \supset D_j$ of D such that $\bigcap_{\alpha \in D_i} \{W_{\alpha\beta_i}\} \neq \emptyset$ for each $1 \leq i \leq j$.

Choose α and γ in D_j. For $1 \leq i \leq j$, choose $x_{\beta_i} \in W_{\alpha\beta_i} \cap W_{\gamma\beta_i}$. For $\beta \in F_\alpha - J$, choose $x_\beta \in W_{\alpha\beta}$. For $\beta \in F_\gamma - J$, choose $x_\beta \in W_{\gamma\beta}$. For $\beta \in B - (F_\alpha \cup F_\gamma)$, choose $x_\beta \in X_\beta$. Then $\{x_\beta\}_{\beta \in B}$ is a point of $U_\alpha \cap U_\gamma$ which is a contradiction.

(8) We next give the partial order form of Martin's axiom and indicate the circle one travels from partial order to Stone space (or Boolean algebra) and back again [36].

Suppose that $(P, \leq)$ is a ccc partially ordered set: that is, every antichain in T is countable. (Why ccc was not called the countable *anti*chain condition no one will ever know.) For $p \in P$, let $p^+ = \{x \in P \mid p \leq x\}$. Give P the partial order topology ($\{p^+ \mid p \in P\}$ forms a basis), and let (P, T) be the resulting (non-Hausdorff) space. Two terms p and q of P are said to be compatible if $p^+ \cap q^+ \neq \emptyset$. A subset X of P is said to be dense and open in $(P, \leq)$ if X is dense and open in (P, T); and X is a filter if, for every finite nonempty subset Y of X, there is a term of X compatible with every term of Y.

Suppose that $|A| < c$ and that $\{X_\alpha\}_{\alpha \in A}$ is a family of dense open sets for $(P, \leq)$. Let Y_α be a maximal family of incompatible terms of X_α. For each $\alpha \in A$, $V_\alpha = \bigcup_{p \in Y_\alpha} (p^+)'$ is dense and open in the Stone space $S(P, T)$. Hence Martin's axiom implies that there is a $v \in \bigcap_{\alpha \in A} V_\alpha$. Let $X = \{p \in P \mid (p^+)' \in v\}$. Since v is a ultrafilter, X is a filter. And if $\alpha \in A$, there is a $p_\alpha \in X \cap Y_\alpha \subset X \cap X_\alpha$.

We therefore have the following:

If $(P, \leq)$ *is a ccc partial order and* $|A| < c$ *and* $\{X_\alpha\}_{\alpha \in A}$ *is a family of dense open sets for* $(P, \leq)$, *then there is a filter* X *in* P *and a subset* $\{p_\alpha\}_{\alpha \in A}$ *of* X *with* $p_\alpha \in X_\alpha$ *for all* α.

It is easy to see how to return through a Stone space from this partial order form of Martin's axiom to the Baire category form.

Let us now prove BAUMGARTNER'S THEOREM: (MA + $\neg$CH) *implies that every Aronszajn tree is special.*

Suppose that $(T, \leq)$ is an Aronszajn tree, $(\mathrm{MA} + \neg\mathrm{CH})$ holds, Q is the set of all rational numbers, and $S = \{\text{finite subsets of } T\}$.

Let P be the set of all functions f from a member of S into Q such that, if x and y belong to the domain of f and $x < y$, then $f(x) < f(y)$.

If f and g belong to P we define $f \leq g$ provided the domain of f is contained in the domain of g and $g \restriction (\text{domain of } f) = f$. Clearly $(P, \leq)$ is a partially ordered set.

For $x \in T$, define $X_x = \{f \in P \mid x \in \text{domain of } P\}$. Clearly X_x is dense and open in P and by assumption $|T| = \omega_1 < c$.

If $(P, \leq)$ is ccc, then by Martin's axiom there is a filter X in P such that, for each $x \in T$, there is a $p_x \in X_x \cap X$. Define a function $f\colon T \to Q$ by $f(x) = p_x(x)$ for all $x \in T$. Since X is a filter, for each $r \in Q$, $\{x \in T \mid f(x) = r\}$ is an antichain in T. Thus T is special.

It remains to prove that $(P, \leq)$ is ccc. Assume, on the contrary, that there is an uncountable subset R of P with each pair of its terms incompatible. By repeatedly taking smaller uncountable subsets of R we can find an uncountable subset $\{f_\alpha\}_{\alpha<\omega_1}$ of R and an integer k and a subset J of T such that:

(1) If $\alpha < \beta < \omega_1$, then $(\text{domain } f_\alpha) \cap (\text{domain } f_\beta) = J$ and $f_\alpha(j) = f_\beta(j)$ for all $j \in J$.

(2) If $\alpha < \omega_1$, (domain of $F_\alpha - J$) has exactly k terms $x_{\alpha 1}, \cdots, x_{\alpha k}$.

(3) If $\alpha < \beta < \omega_1$ and i and j belong to $(1, \cdots, k)$, then the level of the tree containing $x_{\alpha i}$ is less than the level of the tree containing $x_{\beta j}$.

If $1 \leq i \leq k$ and $A \subset \omega_1$ is uncountable, then there is an uncountable subset B of A such that $\{x_{\alpha i}\}_{\alpha \in B}$ is an antichain in T (since otherwise $\{x_{\alpha i}\}_{\alpha \in A}$ would be a Souslin tree which is denied by $(\mathrm{MA} + \neg\mathrm{CH})$. Therefore by repeatedly taking smaller uncountable subsets of ω_1 we find an infinite subset D of ω_1 such that, for all $1 \leq i \leq k$, $\{x_{\alpha i}\}_{\alpha \in D}$ is an antichain in T.

For the pair $(i, j) \subset (1, \cdots, k)$,

(a) if $\alpha < \beta$ and $x_{\alpha i} < x_{\beta j}$, we put (α, β) in pot I,

(b) if $\alpha < \beta$ and $x_{\alpha i} \nless x_{\beta j}$, we put (α, β) in pot II.

By the Erdös-Rado theorem $\omega \to (\omega, \omega)^2$, there is an infinite set $D' \subset D$ such that $D' \subset$ pot II. For suppose that there were three ordinals $\alpha < \beta < \gamma < \omega_1$ all of whose pairs were in pot I. Then $x_{\alpha i} < x_{\gamma j}$ and $x_{\beta i} < x_{\gamma j}$; thus, since T is a tree, $x_{\alpha i} \leq x_{\beta i}$ or $x_{\beta i} \leq x_{\alpha i}$ which contradicts the fact that $x_{\alpha i}$ and $x_{\beta i}$ belong to an antichain.

Repeating the above argument for each pair $(i, j) \subset (1, \cdots, k)$ in turn, we can select $\alpha < \beta$ in D such that $x_{\alpha i} \nless x_{\beta j}$ for any $(i, j) \subset (1, \cdots, k)$. So, by (3), f_α and f_β are compatible.

IV. Martin's Axiom and Normality

Let κ be an infinite cardinal less than c; Martin's axiom implies the following:

[52] (1) If $\{A_\alpha\}_{\alpha<\kappa}$ is a family of subsets of ω with each finite intersection infinite, then there is an infinite subset A of ω such that $A - A_\alpha$ is finite for all $\alpha < \kappa$.

[36] (2) Suppose that A and B are families of cardinality κ of infinite subsets of ω and that $a \cap b$ is finite for all $a \in A$ and $b \in B$. Then there is a subset s of ω such that $b \cap s$ is finite for all $b \in B$ and $a \cap s$ is infinite for all $a \in A$.

(Silver [36]) (3) If X is a cardinality κ subset of the real numbers, then every subset of X is a relative G_δ.

(4) $2^\omega = 2^\kappa$.

If $\omega_1 \leq \kappa < c$, Martin's axiom implies that the following spaces are normal (but not collectionwise normal).

[54] (5) A κ-Cantor tree.

[55] (6) A modified Pixley-Roy space.

[39] (7) An Aronszajn tree.

[53] (8) The square of a κ-Sorgenfrey line.

In the previous chapter we discussed the relationship between Souslin's problem and Martin's axiom. One should expect that (MA + $\neg$CH), specifically tailored to prevent the existence of Souslin lines, might have something to say about generalizations of Souslin's problem. The point of Chapter III was to illustrate this interaction. The point of this chapter is to find broader applications. We frist give four simple combinatorial consequences of Martin's axiom. All of these concern intersection properties for sets of integers and thus have wide influence.

Special areas where we have learned to expect Martin's axiom implications include:

(1) Souslinean problems: that is, problems involving countable density versus countable cellularity with assorted Lindelöf and spread restrictions.

(2) ω_1-tree problems.

(3) ω_1-collectionwise Hausdorff problems.

(4) βN problems.

(5) The preservation of properties in products.

A survey of the area is [56].

The second half of the chapter will use (MA + $\neg$CH) to construct a variety of normal, noncollectionwise normal spaces, spaces which under different set theoertic assumptions fail to be normal. Thus we illustrate the set theoretic nature of many normality problems and the fact that (MA + $\neg$CH) is a real help with these problems.

(1) Let us prove BOOTH'S THEOREM: Suppose $\kappa < c$, $A_\alpha \subset \omega$ for all $\alpha < \kappa$, and $\bigcap_{\alpha \in F} A_\alpha$ is infinite for all finite $F \subset \kappa$. Let $P = \{(f, F) \mid f$ is a finite subset of ω and F is a finite subset of $\kappa\}$. Define $(f, F) \leq (g, G)$ in P provided

$$f \subset g, \quad F \subset G, \quad \text{and} \quad (g - f) \subset \bigcap_{\alpha \in F} A_\alpha.$$

For each $\alpha < \kappa$ and $n \in \omega$ define $X_{\alpha n} = \{(f, F) \in P \mid \alpha \in F$ and $|f| \geq n\}$. Clearly $(P, \leq)$ is a partially ordered set and $X_{\alpha n}$ is dense and open in $(P, \leq)$. We can thus apply the partial order form of Martin's axiom if we prove that $(P, \leq)$ has ccc.

Assume that $\{(k_\alpha, K_\alpha)\}_{\alpha \in L}$ is an uncountable subset of P. There is a finite subset k of ω and an uncountable subset Q of L such that $k_\alpha = k$ for all $\alpha \in Q$. If α and β belong to Q, then $(k_\alpha, K_\alpha) \leq (k, K_\alpha \cup K_\beta)$ and $(k_\beta, K_\beta) \leq (k, K_\alpha \cup K_\beta)$. So every uncountable subset of P has compatible elements; thus $(P, \leq)$ has ccc.

By Martin's axiom we thus have a compatible family $\{f_\alpha, F_\alpha\}_{\alpha<\kappa}$ in $(P, \leq)$ with $(f_\alpha, F_\alpha) \in X_\alpha$ for each $\alpha < \kappa$. Let $A = \bigcup_{\alpha<\kappa} f_\alpha$. We prove that $A - A_\alpha$ is finite for all α. Suppose $a \in A - A_\alpha$. Since $a \in A$, $a \in f_\beta$ for some $\beta < \kappa$. Since $a \notin A_\alpha$ and $\alpha \in F_\alpha$, $a \notin \bigcap_{\gamma \in F_\alpha} A_\gamma$. Since (f_α, F_α) and (f_β, F_β) are compatible, there is a $(g, G) \in P$ such that $(f_\alpha, F_\alpha) \leq (g, G)$ and $(f_\beta, F_\beta) \leq (g, G)$. Since $(f_\beta, F_\beta) \leq (g, G)$ and $a \in f_\beta$, $a \in g$. Since $(f_\alpha, F_\alpha) \leq (g, G)$, $a \in g$, and $a \notin \bigcap_{\gamma \in F_\alpha} A_\gamma$, $a \in f_\alpha$. Thus $(A - A_\alpha) \subset f_\alpha$ and $A - A_\alpha$ is finite, obviously A is infinite.

(2) We now prove SOLOVAY'S THEOREM: (Same song, second verse, $\cdots$.) Suppose that A and B are cardinality κ families of infinite subsets of ω and that $a \cap b$ is finite for all $a \in A$ and $b \in B$. Let $P = \{(f, F) \mid f$ is a finite subset of ω and F is a finite subset of $B\}$. Define $(f, F) \leq (g, G)$ in P provided

$$f \subset g, \quad F \subset G, \quad \text{and} \quad b \cap g \subset f \text{ for all } b \in F.$$

If $(f, F) \leq (g, G)$ and $(g, G) \leq (k, K)$, then $b \in F$ implies that $b \in G$ which implies that $(b \cap k) \subset g$; but $(b \cap g) \subset f$, so $(b \cap k) \subset f$. Thus $(P, \leq)$ is clearly a partially ordered set.

Define $X_{a,n} = \{(f, F) \in P \mid |f \cap a| > n\}$ for all $a \in A$ and $n \in \omega$.

Define $X_b = \{(f, F) \in P \mid b \in F\}$ for all $b \in B$.

If $a \in A$ and $n \in \omega$ and $(k, K) \in P$, since $a - \bigcup K$ is infinite, there is $f \subset (a - \bigcup K)$ with $|f| > n$. Then $(k, K) \leq (f \cup k, K) \in X_{a,n}$ so $X_{a,n}$ is dense (and open) in $(P, \leq)$. If $b \in B$ and $(k, K) \in P$, then $(k, K) \leq (k, K \cup \{b\}) \in X_b$. So each X_b is also dense and open in $(P, <)$.

Assume that $\{k_\alpha, K_\alpha\}_{\alpha \in L}$ is an uncountable subset of P. There is an uncountable subset Q of L and a finite subset k of ω such that $k_\alpha = k$ for all $\alpha \in Q$. For all α and β in Q, $(k_\alpha, K_\alpha) \leq (k, K_\alpha \cup K_\beta)$ and $(k_\beta, K_\beta) \leq (k, K_\alpha \cup K_\beta)$. This shows that $(P, \leq)$ has ccc.

Let $M = (A \times \omega) \cup B$; then $|M| = \kappa$.

Therefore, Martin's axiom implies that there is a compatible family $\{(f_\alpha, F_\alpha)\}_{\alpha \in M}$ in $(P, \leq)$ with $(f_\alpha, F_\alpha) \in X_\alpha$ for all $\alpha \in M$.

Define $s = \bigcup_{\alpha \in M} f_\alpha$.

Suppose that $a \in A$. For every $n \in \omega$, $|f_{a,n} \cap a| > n$. So, for every $n \in \omega$, $|s \cap a| > n$; hence $s \cap a$ is infinite for all $a \in A$.

Suppose that $b \in B$. If $x \in b \cap s$, then $x \in f_\alpha$ for some $\alpha \in M$. Since (f_α, F_α) and (f_b, F_b) are compatible, there is a $(g, G) \in P$ with $(f_\alpha, F_\alpha) \leq (g, G)$ and $(f_b, F_b) \leq (g, G)$. Since $b \in F_b$, $(b \cap g) \subset f_b$. But $f_\alpha \subset g$ so $x \in g$. Thus $x \in f_b$. Since $b \cap s \subset f_b$, $b \cap s$ is finite.

This completes the proof of Solovay's theorem which is typical of Martin's axiom proofs.

(3) At the urging of Frank Tall, Jack Silver observed that Solovay's theorem yields the following:

Suppose that X is a subset of real numbers of cardinality $\kappa < c$, and that $Y \subset X$. There is a countable basis $\mathcal{B} = \{U_i\}_{i \in \omega}$ for the topology of the real line such that, for each pair x and y of real numbers, only finitely many members of $\mathcal{B}$ contain both x and y. For $x \in X$, let $N_x = \{i \in \omega \mid x \in U_i\}$. Let $A = \{N_x\}_{x \in Y}$ and $B = \{N_x\}_{x \in X-Y}$. By Solovay's theorem, assuming Martin's axiom, there is an $s \subset \omega$ such that $s \cap N_x$ is infinite for all $x \in A$ and $s \cap N_x$ is finite for all $x \in X - Y$. For $n \in \omega$ define $V_n = \bigcup\{U_i \mid i \in s \text{ and } i \geq n\}$. Then $Y \subset \bigcap_{n \in \omega} V_n$ but $\bigcap_{n \in \omega} V_n \cap (X - Y) = \emptyset$; hence Y is a G_δ set in X.

(4) Let X be a cardinality κ subset of real numbers for some infinite $\kappa < c$. The number of subsets of X is 2^κ. The number of relative G_δ subsets of X is $2^\omega = c$. So an immediate consequence of (3) is that $2^\omega = 2^\kappa$ if Martin's axiom holds.

(5) Bing and Heath first pointed out [103], [58] that the "separable normal Moore space problem" and the "relative G_δ problem (4)" are the same. We use the Cantor tree to see that this is true.

Suppose that $(B, \leq)$ is a κ-Cantor tree for some $\omega_1 \leq \kappa < c$. Give B the tree topology and suppose that H and K are disjoint closed subsets of B. Let B_k be the kth level of B and without loss of generality we assume that $(H \cup K) \subset B_\omega$. By (3), Martin's axiom implies that H and K are G_δ sets in $H \cup K$. Say $H = (\bigcap_{n \in \omega} U_n) \cap B_\omega$ and $K = (\bigcap_{n \in \omega} V_n) \cap B_\omega$ where $\{U_n | n \in \omega\}$ and $\{V_n | n \in \omega\}$ are decreasing sequences of open sets in the usual line topology. Let $U = \{x \in B \mid$ for some n, x is contained in U_n but x is not contained in $V_n\}$. Then U is open in B, $U \supset H$ and $\overline{U} \cap K = \emptyset$. Hence B is normal. But B is not collectionwise Hausdorff since B_ω is an uncountable closed discrete subset of B whose points cannot be separated by disjoint open sets.

(6) The space we are about to construct is a metacompact, nonmetrizable, Moore space. Martin's axiom implies that this space is normal which is especially interesting since we know that we cannot add both separable and normal to the hypotheses [59].

Let X be a subset of the real line of cardinality κ where $\omega_1 \leq \kappa < c$. If $p \in [X]^2$ and $n \in \omega$, let $U_{p,n} = \{p\}$. If $x \in X$ and $n \in \omega$, let $U_{x,n} = \{x\} \cup \{(x, s) \in [X]^2 \mid |x - s| < 1/n\}$. The space Y obtained by using $\{U_{p,n} \mid p \in X \cup [X]^2$ and $n \in \omega\}$ as a basis for a topology on $X \cup [X]^2$ is called a κ-Pixley-Roy space [59]. (A Pixley-Roy space has many variations.)

Martin's axiom implies that Y is normal; for suppose that H and K are closed and disjoint subsets of Y. Without loss of generality assume that $(H \cup K) = X$. By (3), Martin's axiom implies that $H = (\bigcap_{n \in \omega} U_n) \cap X$ and $K = (\bigcap_{n \in \omega} V_n) \cap X$ where $\{U_n \mid n \in \omega\}$ and $\{V_n \mid n \in \omega\}$ are decreasing sequences of sets open in the usual topology of the line. Let

$$U = H \cup \{(x, s) \in [X]^2 \mid \text{for some } n, [x, s] \subseteq U_n \text{ but } [x, s] \not\subseteq V_n\}.$$

Then U is an open set in Y containing H such that $\overline{U} \cap K = \emptyset$. Thus Martin's axiom and $\kappa < c$ imply that the κ-Pixley-Roy space is normal.

(7) Let us prove FLEISSNER'S THEOREM: (MA + $\neg$CH) *implies that Aronszajn trees are normal.*

Let $(T, \leq)$ be an Aronszajn tree (with the induced tree topology). From Baumgartner's theorem in Chapter III(8), we know that T is special. Thus $T = \bigcup_{n \in \omega} A_n$ where A_n is an antichain.

Let L_n be the set of all levels of T which have nonempty intersection with A_n. There is an n for which L_n is stationary (meets every uncountable closed subset of ω_1). If Λ is a subset of an ordinal λ, a function $f\colon \Lambda \to \lambda$ such that $f(\alpha) < \alpha$ for all $\alpha \in \Lambda$ is called a pressing down function. It is well known that any pressing down function on a stationary set is constant on a stationary set. So there do not exist disjoint open sets separating the points of the discrete closed set A_n. Hence T is not collectionwise normal (or metrizable).

Jones knew that a special Aronszajn tree was a nonmetrizable Moore space and he speculated that there might be a normal one [42], [60]. So the problem of the normality of a special Aronszajn tree became known as the *Jones road problem.* Jones inserted intervals between successive elements in the tree to make it connected and hence a "road" but it does not change the problem.

Let us prove that (MA + $\neg$CH) implies that T is normal. Assume that H and K are closed and disjoint in T.

For $t \in T$ and $n \in \omega$, let

$$U(t, n) = \{x \leq t \mid \{y \in T \mid x \leq y < t\} \cap \textstyle\bigcup_{k \leq n} A_k = \emptyset\}.$$

Observe that $\{U(t, n)\}_{n \in \omega}$ is a local basis at t.

Let P be the set of all functions f from a finite subset of $H \cup K$ into ω such that, for all $x \in (\text{domain } f) \cap H$ and $y \in (\text{domain } f) \cap K$, $U(x, f(x)) \cap U(y, f(y)) = \emptyset$. Define $f \leq g$ in P provided $(\text{domain } f) \subset (\text{domain } g)$ and $g \restriction (\text{domain } f) = f$. Then $(P, \leq)$ is a partially ordered set. By an argument similar to the one given in Chapter III(8), $(P, \leq)$ has ccc.

For $t \in H \cup K$, define $X_t = \{f \in P \mid t \in (\text{domain } f)\}$. Then X_t is open and dense in $(P, \leq)$.

(MA + $\neg$CH) implies that there is a family $\{f_t\}_{t \in H \cup K}$ of compatible members of P such that $f_t \in X_t$ for each $t \in H \cup K$. Let $U = \bigcup_{t \in H} U(t, f_t(t))$ and $V =$

$\bigcup_{t \in K} U(t, f_t(t))$. Then U and V are open, disjoint covers of H and K, respectively. Hence T is normal.

(8) The same type of proof, now so familiar, gives the following example. Again we know of no "real" paracompact spaces whose product is normal but not paracompact.

Let us prove PRZYMUSIŃSKI'S THEOREM: (MA + $\neg$CH) *and* $\kappa < c$ *imply that the* κ-*Sorgenfrey square is normal.*

The Sorgenfrey line [12] is the real line with the set of all intervals $[a, b)$ for $a < b$ used as a basis for the topology. It was originally constructed as an example of a paracompact space whose square is not normal. A subspace S of the Sorgenfrey line of cardinality $\kappa < c$ is called a κ-Sorgenfrey line.

Observe that $S \times S$ is not collectionwise Hausdorff since a diagonal of slope -1 is a discrete closed set whose points cannot be separated by disjoint open sets. However $S \times S$ is normal. Assume that H and K are disjoint closed subsets of $S \times S$. $\overline{U}$ will denote the ordinary closure of U in the plane.

For $(s, t) \in S \times S$ and $n \in \omega$ define

$$U_n(s, t) = [s, s + 1/n) \times [t, t + 1/n).$$

For $x \in H$ let

$$J_x = \{n \in \omega \mid \overline{U_n(x)} \cap K = \emptyset\}$$

and for $x \in K$, let

$$J_x = \{n \in \omega \mid \overline{U_n(x)} \cap H = \emptyset\}.$$

Let P be the set of all functions f from a finite subset of $H \cup K$ into ω such that $x \in (\text{domain } f) \cap H$ and $y \in (\text{domain } f) \cap K$ imply that $f(x) \in J_x$, $f(y) \in J_y$, and $U_{f(x)}(x) \cap U_{f(y)}(y) = \emptyset$.

Define $f \leq g$ in P provided $(\text{domain } f) \subset (\text{domain } g)$ and $g \restriction (\text{domain } f) = f$. Then $(P, \leq)$ is a partially ordered set.

For $x \in H \cup K$, define $H_x = \{f \in P \mid x \in (\text{domain } f)\}$. Since each X_x is open and dense, if $(P, \leq)$ has ccc, Martin's axiom implies that there is a family $\{f_x\}_{x \in H \cup K}$ of comparable terms of P such that $f_x \in X_x$ for all $x \in H \cup K$. Then $U = \bigcup_{x \in H} U_{f(x)}(x)$ and $V = \bigcup_{x \in K} U_{f(x)}(x)$ are disjoint open subsets of $S \times S$ containing H and K, respectively.

Thus $S \times S$ is normal, assuming (MA + $\neg$CH), if $(P, \leq)$ has ccc. Assume that Y is an uncountable subset of P. There is an uncountable subset Z of Y such that $f \in Z$ implies that the domain of f has precisely i terms $f_1, \cdots, f_i$ belonging to H and j terms $f_{i+1}, \cdots, f_{i+j}$ belonging to K, and, for $1 \leq n \leq i + 1$ and f and g in Z, $f(f_n) = g(g_n)$. Let $\{V_k\}_{k \in \omega}$ be a countable basis for the ordinary topology of the plane. For each f in Z, $n \in (1, \cdots, i)$, and $m \in (i + 1, \cdots, i + j)$, there is a pair h and k of integers such that $f_n \in V_h$, $f_m \in V_k$, and, if $x \in V_h$ and $y \in V_k$, then $\overline{U_{f(f_n)}(x) \cap U_{f(f_m)}(y)} = \emptyset$.

By repeated choice of n and m, we can choose an uncountable subset W of Z such that for all f and g in W, $n \in (1, \cdots, i)$ and $m \in (i+1, \cdots, i+j)$, $\overline{U_{f(f_n)}(f_n)} \cap \overline{U_{g(g_m)}(g_m)} = \emptyset$. Hence all of the terms of W are compatible.

V. Hereditary Separability and Hereditary Lindelöfness

[61] (1) There is a nonregular S-space.

[61] (2) There is a nonregular L-space.

[62] (3) There is a nonregular space of countable spread which is not the union of a hereditarily Lindelöf subspace and a hereditarily separable subspace.

[63] (4) There is a machine for reducing nonnormality to noncomplete regularity as well as a machine for nonnormality.

(5) A Souslin line is a regular L-space.

[64], [65] (6) A Souslin line implies that there is a regular (normal or not normal) S-space.

[66], [67] (7) CH implies that there is a regular S-space in 2^c (of cardinality c^+ with forcing).

[66], [67] (8) CH implies that there is an L-space of cardinality c such that every countable subset is discrete and every uncountable subset has weight 2^c.

The contrast in topology between limit point properties and covering properties is a fundamental one. Limit point properties are more intuitive and easy to recognize. On the other hand the beautiful theorems are true for covering properties. The property that every open cover has a finite subcover is clearly the more basic compactness property rather than that every infinite set has a limit point.

Separability and Lindelöfness are two such contrasting properties which are equivalent for metric spaces. The lexicographically ordered square is a good example of a compact space which is not separable; $\beta N - N$ is another. The Cantor tree is a good example of a separable space which is not Lindelöf; βN minus a point of $\beta N - N$ is another. However all of these spaces have uncountable discrete subspaces which are, of course, neither separable nor Lindelöf.

General topologists have been becoming increasingly interested in the relationship between *hereditarily* separability and *hereditarily* Lindelöfness. Let us call a space an S-space if it is hereditarily separable but not Lindelöf and an L-space if it is hereditarily Lindelöf but not separable.

We know that there are nonregular (but Hausdorff) [61] S-spaces and L-spaces. But we know of no "real" regular S-spaces or L-spaces. CH or there exists a Souslin line implies that there are regular S-spaces and regular L-spaces. The strongest known L-space is probably the Souslin line; the strongest known S-space is probably the one constructed using $V = L$ ([69] in Chapter VI(8)) which is perfectly normal, locally compact, first countable, and countably compact. There are several obvious questions:

(a) Is there, without set theoretic assumptions, an S-space or an L-space?

(b) Does (MA + $\neg$CH) imply that there is no regular S-space or L-space?

(c) Is the existence of a regular L-space equivalent to the existence of a regular S-space?

(d) If a regular space has no uncountable discrete subset, is it the union of a hereditarily separable subspace and a hereditarily Lindelöf subspace? (A nonregular example is known [62].)

(e) Is there a "real" regular S-space of cardinality 2^c? Every first countable Lindelöf space has cardinality $\leq c$ (see Chapter II(8)) as does every hereditarily Lindelöf regular space. We have shown (Chapter II(7)) that every space of cardinality greater than 2^c has an uncountable discrete subspace. Juhász and Hajnal prove that it is consistent with the usual axioms for set theory and CH, that there be a normal subset of 2^c of cardinality c^+ which is an S-space.

(f) Is there a "real" S-space which is regular but not normal? All of the first examples of regular S-spaces turned out to be normal by accident [64], [66]. And every regular Lindelöf space is normal. However, Burton Jones has shown that a nonnormal S-space can be used to construct one which is not completely regular [63]. And there is also a technique [65] for eliminating normality from some S-spaces.

Common factors in all of these problems are the fact that regularity is the difficult-to-achieve separation axiom and the fact that we have no idea what consequences (MA + $\neg$CH) yields. In all cases there may, of course, be "real" examples.

(1) Let (X, T) be any hereditarily separable, hereditarily Lindelöf (Hausdorff) topological space of cardinality ω_1: a cardinality ω_1 subset of the line, for instance. Minimally well order $X = \{x_\alpha\}_{\alpha<\omega_1}$. Let Y and Z be the spaces whose underlying set is X and whose topologies are generated by the subbases $T \cup \{\{x_\beta \mid \beta \leq \alpha\} \mid \alpha < \omega_1\}$ and $T \cup \{\{x_\beta \mid \alpha \leq \beta\} \mid \alpha < \omega_1\}$, respectively.

The space Y is hereditarily separable. For suppose that $\{x_\alpha\}_{\alpha\in I} \subset X$. There is a countable $J \subset I$ such that $\{x_\alpha\}_{\alpha\in J}$ is dense in (X, T). Choose $\beta \in \omega_1$ such that $\alpha < \beta$ for all $\alpha \in J$. Then $\{x_\alpha \mid \alpha \in I$ and $\alpha < \beta\}$ is dense in $\{x_\alpha\}_{\alpha\in I}$ in Y. Hence Y is hereditarily separable, and Y is an S-space.

(2) The space Z is not separable but we prove that it is hereditarily Lindelöf and thus an L-space.

Suppose that $W \subset X$, $\{U_\alpha\}_{\alpha\in I}$ covers W, and each $U_\alpha = W \cap (V_\alpha \cap W_\alpha)$ where $V_\alpha \in T$ and, for some $\gamma_\alpha \in \omega_1$, $W_\alpha = \{x_\beta \in X \mid \beta \geq \gamma_\alpha\}$. Such U_α are basic sets for the topology of W in Z. There is a countable subset J of I such that $\{V_\alpha\}_{\alpha\in J}$ covers W. Choose $\beta \in \omega_1$ such that $\gamma_\alpha < \beta$ for all $\alpha \in J$. Then $\{U_\alpha\}_{\alpha\in J}$ covers $\{x_\delta \in W \mid \delta > \beta\}$. Since $\{x_\delta \in W \mid \delta \leq \beta\}$ is countable, countably many U_α cover W and Z is thus proved hereditarily Lindelöf.

(3) In both of the preceding problems as well as in the one which follows, you observe that the "induced" topology is finer than the "usual" topology and therefore the spaces constructed are Hausdorff; but regularity, even when present in the original spaces, is destroyed.

We construct a space of countable spread which is not the union of hereditarily separable and hereditarily Lindelöf subspaces. We use the space X from (1).

Let $G = \{(U, V, \gamma, \lambda, F) \mid U$ and V are open in X, γ and λ belong to ω_1 and F is a finite subset of $\omega_1\}$. Let $O_{(U,V,\gamma,\lambda,F)} = \{(x_\alpha, x_\beta) \in X \times X \mid \alpha \leq \gamma, x_\alpha \in U, x_\beta \in V$ and, if $\alpha \in F$, then $\beta \geq \lambda\}$. The desired space $S = X \times X$ with the topology induced by taking $\{O_g\}_{g \in G}$ as a basis.

Let us prove that S has countable spread. Assume that, on the contrary, D is an uncountable discrete subset of S. For each $\alpha < \omega_1$, the subspace $\{x_\alpha\} \times X$ of S is homeomorphic to Z which is hereditarily Lindelöf, so $D \cap (\{x_\alpha\} \times X)$ is countable. Since D is uncountable, there is an uncountable subset I of ω_1 and, for each $\alpha \in I$, a $\gamma_\alpha \in \omega_1$ such that $(x_\alpha, x_{\gamma_\alpha}) \in D$. Since $X \times X$ is hereditarily separable in the usual topology, there is a countable subset J of I such that $\{(x_\alpha, x_{\gamma_\alpha}) \mid \alpha \in J\}$ is dense in $\{(x_\alpha, x_{\gamma_\alpha}) \mid \alpha \in I\}$ in the usual topology of $X \times X$. But then, if $\beta > \alpha$ for all $\alpha \in J$, $(x_\beta, x_{\gamma_\beta})$ is a limit point of $\{(x_\alpha, x_{\gamma_\alpha}) \mid \alpha \in I\}$ which contradicts the fact that D is discrete. Thus S has countable spread.

Now, in order to prove that it is impossible, let us assume that $S = A \cup B$ where A is herediatrily separable and B is hereditarily Lindelöf. Since no uncountable subset of Z is separable, $A \cap (\{x_\alpha\} \times X)$ is countable for each $\alpha < \omega_1$. So, for each $\alpha < \omega_1$, there is a γ_α such that $(x_\alpha, x_{\gamma_\alpha}) \in B$. Define $O_\alpha = \bigcup_{\beta \leq \alpha}(\{x_\beta\} \times X)$. Then $\{O_\alpha \cap B\}_{\alpha < \omega_1}$ is an open cover of $\{(x_\alpha, x_{\gamma_\alpha})\}_{\alpha \in \omega_1}$ with no countable subcover so B is not hereditarily Lindelöf.

(4) This machine was proposed by Burton Jones who had worked on hereditary separability problems in the 1930's.

Suppose that X is a space which is not normal and that A and B are disjoint closed subsets of X which cannot be separated. For each $n \in \omega$, let X^n be a copy of X. If $x \in X$, x^n will be the copy of x in X^n and, if $K \subset X$, K^n will be the copy of K in X^n.

Let Y be the disjoint union of all X^n, and let Z be the quotient space of Y obtained by identifying B^{2n} with B^{2n+1} and A^{2n+1} with A^{2n+2}. Choose a point p not in any X^n and our final space is $W = Z \cup \{p\}$ where O is open if and only if

(a) $O \cap Z$ is open in Z and

(b) there is an $n \in \omega$ with $\bigcup_{m>n} X^m \subset O$ if $p \in O$.

Properties such as regularity, hereditary separability, hereditary Lindelöfness and first countability are preserved in W when found in X. But W is *not* completely regular since A^0 and $\{p\}$ cannot be separated in W. Therefore, if there is an example of a regular S-space which is not normal, there is also an example of one which is not completely regular.

We also have a machine for getting rid of normality [65]. Assume that X is a regular S-space with an open basis of closed countable sets. The example in (6) and Chapter VI(8), for example, are such spaces. Take two copies X_1 and X_2 of X. Choose an uncountable subset A of the Cantor set such that every countable subset of A is a relative G_δ. Let C be the Cantor tree which has A as its ωth level. Pick disjoint subsets B and D of A which cannot be separated in C. Such sets exist if $2^\omega < 2^{\omega_1}$. Assume that $|X| = \omega_1$ and let $f: B \to X_1$ and $g: D \to X_2$ be one-to-one correspondences. We make the identification space and use the identification topology induced by f and g on the disjoint union of X_1, X_2, and C. The resulting space is a nonnormal S-space.

(5) A Souslin line is a perfectly normal, locally compact L-space.

(6) We now use the existence of a Souslin line to construct a regular S-space. Let $(T, \leq)$ be a Souslin tree. Let T_α be the αth level of T and, if $t \in T$, let $l(t)$ be the level of the tree to which t belongs.

Suppose that $H \subset T$. Let $R = \{r \in T \mid r < h \in H\}$. Let Q be a maximal antichain in $T - R$ and choose $\alpha \in \omega_1$ with $l(t) < \alpha$ for all $t \in Q$. If $t \in R \cap T_\alpha$, a maximal antichain in $\{x \in H \mid x > t\}$ is a maximal antichain in $\{x \in T \mid x > t\}$. If $\gamma \in \omega_1$ and $t \in R \cap T_\gamma$, define $\gamma = \gamma_0 < \gamma_1 < \cdots$ and define antichains $A_0, A_1, \cdots$ by induction such that A_n is a maximal antichain in $\{x \in H \cap (\bigcup_{\beta > \gamma_n} T_\beta) \mid x > t\}$ and $\gamma_{n+1} > l(x)$ for all $x \in A_n$. Then, if γ' is the limit of $\gamma_0, \gamma_1, \cdots$, the tail of every chain running through the tree from t to level γ' hits infinitely many A_n.

In order to make T into an S-space, for each $t \in T$ let $\mathcal{A} = \{(n, \alpha, t) \in \omega \times \omega_1 \times T \mid \alpha$ is a limit level $< l(t)\}$. If α is a limit ordinal in ω_1 select $\alpha^0 < \alpha^1 < \cdots$ having α as a limit.

For each $A = (n, \alpha, t) \in \mathcal{A}$, we choose a chain $Z(A)$ in T running from the predecessor of t in T_{α^n} to T_α such that $Z(A)$ is not contained in $\{p < r\}$ for any $r \in T_\alpha$. We make the choice in such a way that, for any $\gamma < \omega_1$, there is a tail $Z_\gamma(A)$ of $Z(A)$ such that $Z_\gamma(A) \cap Z_\gamma(m, \beta, r) = \emptyset$ unless the same term of T_γ precedes both t and r.

Topologize T by declaring a subset V of T to be open if, for each $t \in V$ and limit $\alpha < l(t)$, there is a $k \in \omega$ such that, for all $n > k$, a tail of $Z(n, \alpha, t)$ is contained in V.

To see that T with this topology is not Lindelöf just observe that $\{\bigcup_{\beta<\alpha} T_\beta\}_{\alpha<\omega_1}$ is an open cover of T and certainly has no countable subcover. The argument at the beginning of our discussion shows that T is hereditarily separable. Trivially T is T_1. We prove that T is regular by proving that is is normal!

Suppose that H and K are disjoint and closed. By the argument at the beginning, there is a $\gamma < \omega_1$ such that, for $t \in T_\gamma$, either $\{x > t\} \subset H$ or $\{x > t\} \subset K$ or $\{x > t\} \cap (H \cap K) = \emptyset$. We can thus assume that $\bigcup_{\delta > \gamma} T_\delta \subset (H \cup K)$ and that γ is a nonlimit ordinal. If $t \in T$ and $\alpha < l(t)$ is a limit, let $A_n = (n, \alpha, t)$ for each $n \in \omega$; then there is a $k \in \omega$ and, for every $n > k$, a tail $Y(A_n)$ of $Z(A_n)$ which intersects H only if $t \in H$ and intersects K only if $t \in K$; take $Y(A_n) \subset Z_\gamma(A_n)$. Let $Y_t = \bigcup_{n>k} Y(A_n)$. Then $\bigcup_{t \in H} Y_t$ and $\bigcup_{t \in K} Y_t$ are disjoint closed sets containing H and K, respectively. We repeat this process infinitely many times and thus develop disjoint open sets containing H and K, respectively; hence T is normal and regular.

(7) Juhász and Hajnal use CH to construct a regular S-space in 2^c as well as a

regular L-space. As was true in (6), the hard part of the proof is really the combinatorics of selecting sets with the right intersection properties and again we avoid this problem.

If $A \subset c$, let D_A be the set of all functions whose domain is a finite subset of A and whose range is 2.

If $Z \subset 2^c$ we say that Z is *dense in a tail* if there is a $\beta < c$ such that, for every $d \in D_{c-\beta}$, there is an $f \in Z$ with $f \restriction (\text{domain } d) = d$. (Here Z is dense in the β-tail.)

Juhász and Hajnal prove that CH implies:

(H) *There is an $X \subset 2^c$ such that $|X| = c$ and every infinite subset of X is dense in a tail.*

Suppose CH and that X satisfying (H) has been chosen. Let W be the set of all members of X which are not zero from some point on; clearly $X - W$ is finite. If $W = \{f_\alpha\}_{\alpha<c}$, define $g_\alpha \in 2^c$ by $g_\alpha(\beta) = 0$ if $\beta \leq \alpha$ and $g_\alpha(\beta) = f_\alpha(\beta)$ if $\beta > \alpha$. Let $Y = \{g_\alpha\}_{\alpha<c}$; we prove that Y is an S-space.

Observe that every infinite subset of $\{g_\alpha\}_{\alpha<c}$ is dense in a tail. For suppose that M is an infinite subset of c. There is a $\beta < c$ such that $\{f_\alpha\}_{\alpha\in M}$ is dense in the β-tail. Choose $\gamma > \{\beta\} \cup M$; then $\{g_\alpha\}_{\alpha \in M}$ is dense in the γ-tail.

We see that Y is not Lindelöf because, if for each $\alpha < c$ we define $U_\alpha = \{f \in Y \mid f(\alpha) = 1\}$, then $\{U_\alpha\}_{\alpha<c}$ has no countable subcover.

But Y is hereditarily separable. Assume that $Z \subset Y$ is not separable. There is $\{z_\alpha\}_{\alpha<\omega_1} \subset Z$ such that $z_\alpha \notin \overline{\{z_\beta\}_{\beta<\alpha}}$. Thus, for all $\alpha < \omega_1$, there is a finite subset F_α of ω_1 such that $z_\alpha \restriction F_\alpha \neq z_\beta \restriction F_\alpha$ for any $\beta < \alpha$.

Using partition calculus there is an uncountable subset M of ω_1 and an $F \subset \omega_1$ such that $F_\alpha \cap F_\beta = F$ and $z_\alpha \restriction F = z_\beta \restriction F$ for all α and β in M. Choose a countable infinite $A \subset M$. Then there is a $\beta < \omega_1$ such that $\{z_\alpha\}_{\alpha\in A}$ is dense in the β-tail. Choose $\gamma > A$ such that $F_\gamma - F > \beta$ (this is possible since $\omega_1 = c$). Since $\{z_\alpha\}_{\alpha \in A}$ is dense in the β-tail, there is an $\alpha \in A$ such that $z_\alpha \restriction (F_\gamma - F) = z_\gamma \restriction (F_\gamma - F)$; so $z_\alpha \restriction F_\gamma = z_\gamma \restriction F_\gamma$. But this contradicts $\alpha < \gamma$ in the definition of F_γ. Hence Y is hereditarily separable and an S-space; Y is obviously completely regular and is actually normal.

(8) Again assume CH. Our idea is to construct a special subset of 2^c of cardinality c, and then to build a space whose open sets correspond to the points from (7) and whose points correspond to the open sets from (7).

If $f \in 2^c$, define $f' \in 2^c$ by $f'(\alpha) = 0$ if and only if $f(\alpha) = 1$. If $Z \subset 2^c$, we say that Z is *compatible* if $f \in Z$ implies that $f' \notin Z$. We say that a family G of subsets of 2^c *covers a tail* if there is a $\beta < c$ such that, for all $\alpha > \beta$, there is a $Z \in G$ with $f(\alpha) = 1$ for all $f \in Z$.

Using CH, Juhász and Hajnal prove

(J) *There is $X = (\{f_\alpha\}_{\alpha<c} \cup \{f_{\alpha'}\}_{\alpha<c}) \subset 2^c$ such that, if G is any infinite family of disjoint, compatible, finite, nonempty subsets of X, then G covers a tail.*

If X satisfies (J) and $\alpha \in c$, define $g_\alpha \in 2^c$ by $g_\alpha(\beta) = f_\alpha(\beta)$ for $\beta > \alpha$, $g_\alpha(\alpha) = 1$, and $g_\alpha(\beta) = 0$ for all $\beta < \alpha$. Let $Y = (\{g_\alpha\}_{\alpha<c} \cup \{g_{\alpha'}\}_{\alpha<c})$; observe that Y satisfies (J) also.

For each $g \in Y$, define $U_g = \{\alpha \in c \mid g(\alpha) = 1\}$. Let L be c topologized by using $\{U_g\}_{g\in Y}$ as a subbasis.

Since $\alpha \in U_{g\alpha}$, L is indeed a topological space. Since $(U_{g\alpha}, U_{g\alpha'})$ is an open cover of L by disjoint sets, L is Hausdorff, 0-dimensional, and regular. If $\alpha < c$, α is not a limit point of $\{\beta < \alpha\}$, so L is not separable.

We complete the proof that L is an L-space by showing that L is hereditarily Lindelöf. Assume on the contrary that $\{V_\alpha\}_{\alpha<c}$ is a family of basic open sets such that each V_α contains a point not in $\bigcup_{\beta<\alpha} V_\beta$. By a delta-system argument, there is an uncountable $A < c$ and an n and k in ω such that (a) for all $\alpha \in A$, V_α is the intersection of n subbasic open sets $V_{\alpha 1}, \cdots, V_{\alpha n}$ and (b) for all $\alpha \neq \beta$ in A, $V_{\alpha i} = V_{\beta i}$ for all $i < k$, and $V_{\alpha i} \neq V_{\beta j}$ for all j and $i \geq k$. Let $Z_\alpha = \{g \in Y \mid V_{\alpha i} = U_g$ for some $i \geq k\}$. Choose a countable infinite subset B of A, and let $G = \{Z_\alpha\}_{\alpha\in B}$. The members of G are disjoint, finite, nonempty, and compatible; so G covers a tail. Hence there is a $\beta < c$ such that $\{V_\alpha\}_{\alpha\in B}$ covers $\{\alpha > \beta\}$ which contradicts the definition of $\{V_\alpha\}_{\alpha<c}$.

With forcing the weight can be increased [65].

Judy Roitman [62] uses these spaces to show that CH implies that there is a subspace of 2^c of cardinality c which is hereditarily separable and hereditarily Lindelöf but every subspace of cardinality c has weight c.

Another related theorem (found in [62]) due to Silver is that (MA + $\neg$CH) implies that, if A is an infinite subset of 2^{ω_1}, there is an infinite subset of A not dense in any tail. Silver uses the combinatorial theorem: (MA + $\neg$CH) implies that if E is a countable infinite family of uncountable subsets of ω_1, then there is an infinite $D \subset E$ such that either $\bigcap_{A\in D} A$ is uncountable or $\bigcap_{A\in D} (\omega_1 - A)$ is uncountable.

All of the theorems in this chapter have α-separable and α-Lindelöf generalizations with similar proofs.

VI. Gödel's Constructible Universe

[70] (1) $\lozenge$ implies $\lozenge_1$ implies $\lozenge_2$ implies $\lozenge$.

[69] (2) $\lozenge$ implies $\clubsuit$.

[70] (3) $\lozenge$ implies CH.

[70] (4) $\lozenge$ implies there is a Souslin line.

[39] (5) $\lozenge$ implies no special Aronszajn tree is normal.

[72] (6) $V = L$ implies every normal space of character $\leq c$ is collectionwise Hausdorff.

[68] (7) CH implies there is a first countable S-space.

[69] CH + $\clubsuit$ imply there is a Hausdorff topology T on ω_1 which is:

(a) perfectly normal and hereditarily normal,

(b) countably compact,

(c) hereditarily separable,

(d) first countable,

(e) there is a basis of compact countable open and closed sets,

(f) open sets are countable or cocountable,

(g) for every $\alpha < \omega_1$, α is open;

hence (ω_1, T) is neither compact nor Lindelöf.

General topologists continue to be amazed at the variety of consequences of (MA + $\neg$CH), but it is the contrast between these consequences and the consequences of CH that are so striking. However, there is a model for set theory which has been well known for 40 years which accentuates this contrast: Gödel's constructible universe. This model is commonly known as L, the smallest model for set theory; it includes no set that is not absolutely required to be there by the axioms. Almost all mathematicians know that the generalized continuum hypothesis, GCH, holds in L and that GCH is thus consistent with the usual axioms for set theory. It is common practice to make this assumption. However, it is only rather recently that the fine structure of L has begun to be studied. The basic paper in the area is by R. B. Jensen [70]; K. J. Devlin has also been very active [71]. All sorts of combinatorial and partial order consequences of assuming that $V = L$ have been investigated; $\lozenge$ was first suggested by Jensen and this chapter is a small monument to the consequences of $\lozenge$. $\lozenge$ is a simple combinatorial statement which holds if $V = L$; among the consequences of $\lozenge$ are CH and the existence of a Souslin line and thus essentially all of the theorems from the previous chapter. We will prove others in the same vein in this chapter.

Before beginning let us mention two facts. First of all, $\Diamond$ as presented here says something about ω_1 and its subsets. Similar versions of $\Diamond$ exist for all regular cardinals κ and are true in L just as GCH as well as CH holds in L. Also it should be mentioned that $\Diamond$ in no way exhausts the possibilities for combinatorial consequences of assuming that $V = L$. In the proof of (8) Fleissner uses a stronger "$\Diamond$-like" (but more complicated) consequence of $V = L$.

One reason that the assumption $V = L$ has so many consequences in topology is that it offers great leeway in the construction of trees. Not only does $\Diamond$ imply that there are Souslin trees; $\Diamond$ also implies that there are Aronszajn trees which are neither Souslin nor special [71]. In L there are Souslin trees of cardinality κ for all cardinals κ which are the successors of regular cardinals. Jensen has proved that in L there are Souslin lines S satisfying (a) $A_1 + A_3 + R$, (b) $A_2 + A_3 + R$, (c) $A_2 + A_3 + \neg R$, (d) $A_4 + R$, or (e) $A_4 + \neg R$ where:

R = S is isomorphic to S^{-1}.

A_1 = S has exactly 2^c automorphisms.

A_2 = S has exactly c automorphisms.

A_3 = any two open intervals in S are isomorphic.

A_4 = no two distinct open intervals in S are isomorphic.

After playing with $\Diamond$ for a while one even sees why such variety should be expected.

(1) Jensen's $\Diamond$ [70] can take any of the following forms. Recall that a subset of ω_1 is *stationary* if it meets every closed unbounded subset of ω_1.

$\Diamond$ = there is a family $\{f_\alpha\}_{\alpha<\omega_1}$ of functions such that f_α maps α into α and, if f maps ω_1 into ω_1, then $\{\alpha \mid f \restriction \alpha = f_\alpha\}$ is stationary.

$\Diamond_1$ = there is a family $\{S_\alpha\}_{\alpha<\omega_1}$ of subsets of ω_1 such that $S_\alpha \subset \alpha$ and, if $S \subset \omega_1$, then $\{\alpha \mid S \cap \alpha = S_\alpha\}$ is stationary.

$\Diamond_2$ = there is a family $\{M_\alpha\}_{\alpha<\omega_1}$ of subsets of $\omega_1 \times \omega_1$ such that $M_\alpha \subset \alpha \times \alpha$ and, if if $M \subset \omega_1 \times \omega_1$, then $\{\alpha \mid M \cap (\alpha \times \alpha) = M_\alpha\}$ is stationary.

$\Diamond$ implies $\Diamond_1$ for we can let S_α be the range of f_α.

$\Diamond_1$ implies $\Diamond_2$ for there is a one-to-one function from ω_1 onto $\omega_1 \times \omega_1$ which maps α onto $\alpha \times \alpha$ for all limit ordinals α.

$\Diamond_2$ implies $\Diamond$ for let $A = \{\alpha \in \omega_1 \mid M_\alpha$ is a function from α into $\alpha\}$. Define $f_\alpha = M_\alpha$ for all $\alpha \in A$ and define f_α arbitrarily otherwise.

(2) Ostaszewski defines $\clubsuit$ as follows. Let $\{\lambda_\alpha\}_{\alpha<\omega_1}$ be the order preserving indexing of the limit ordinals in ω_1.

$\clubsuit$ = there is a family $\{S_\alpha\}_{\alpha<\omega_1}$ of subsets of ω_1 such that S_α is a cofinal subset of λ_α and, if S is an uncountable subset of ω_1, then there is an $\alpha \in \omega_1$ with $S_\alpha \subset S$.

In order to prove that $\Diamond$ *implies* $\clubsuit$ assume that $\{S'_\alpha\}_{\alpha\in\omega_1}$ satisfies the conditions for $\{S_\alpha\}_{\alpha\in\omega_1}$ in $\Diamond_1$. Let $A = \{\alpha < \omega_1 \mid S'_{\lambda_\alpha}$ is cofinal in $\lambda_\alpha\}$. Define $S_\alpha = S'_{\lambda_\alpha}$ for all $\alpha \in A$ and otherwise define $S_\alpha = \lambda_\alpha$. Then $\clubsuit$ is satisfied for suppose S is an uncountable subset of ω_1. Let S^* be the closed unbounded set consisting of all limit points of S in ω_1. By $\Diamond$ there is an $\alpha \in S^*$ such that $S'_\alpha = S \cap \alpha$. But $\alpha = \lambda_\beta$ for some β and S'_α is cofinal in α, so $S_\beta \subset S$. Hence $\clubsuit$ is satisfied.

(3) ♣ and (MA + ¬CH) are contradictory. In fact Devlin has proved that (♣ + CH) is equivalent to ◊. We now prove that ◊ implies CH. Let $\{S_\alpha\}_{\alpha<\omega_1}$ be a family satisfying ◊. If X is a subset of ω, let α_X be the smallest infinite ordinal α such that $S_\alpha = X$; by ◊ there is one. If X and Y are different subsets of ω, $\alpha_X \neq \alpha_Y$; hence $c \leq \omega_1$.

(4) We now prove that ◊ *implies there is a Souslin tree*; the ease of this pretty proof will surely be lost in the notation. Some preliminaries are needed.

If X is an infinite countable set we can choose an infinite subset X^0 of X and construct a tree $T_X = (X, \leq)$ of height ω such that (a) X^0 is the first level of T_X and (b) if x belongs to the nth level of T_X there are exactly two elements of the $(n+1)$st level of T_X preceded by x.

If $(A, \leq)$ and $(B, \leq)$ are trees then, by $(A, \leq) + (B, \leq)$, we mean $A \cup B$ with the orders in A and B preserved and, if $a \in A$ and $b \in B$, then $a \leq b$ if there is an $x \in A \cap B$ with $a \leq x \leq b$. If $(A, \leq)$ is a subtree of $(B, \leq)$ or if the last level of $(A, \leq)$ is the first level of $(B, \leq)$ this is a tree; in general, of course, it is not even a partial order.

Let $\{\lambda_\alpha\}_{\alpha<\omega_1}$ be the order preserving indexing of the limit ordinals in ω_1.

Assume that $\{S_\alpha\}_{\alpha<\omega_1}$ is a family satisfying ◊.

We construct a Souslin tree by induction. For each $\alpha < \omega_1$ we construct a tree $(\lambda_\alpha, \leq)$ of height λ_α such that:

(1) if $\beta < \alpha$ and $\delta < \lambda_\beta$, then the δth level of $(\lambda_\beta, \leq)$ is the δth level of $(\lambda_\alpha, \leq)$ and, if $x < y$ in $(\lambda_\beta, \leq)$ then $x < y$ in $(\lambda_\alpha, \leq)$.

(2) If $\beta < \lambda_\alpha$ and x belongs to the βth level of $(\lambda_\alpha, \leq)$, then there are at least two elements of $(\lambda_\alpha, \leq)$ which follow x.

Define $(\lambda_0, \leq) = T_\omega$.

Suppose that $\gamma < \omega_1$ and that $(\lambda_\alpha, \leq)$ has been defined for all $\alpha < \gamma$ satisfying (1) and (2) above.

If γ is a limit ordinal, then $(\lambda_\gamma, \leq) = \Sigma_{\alpha<\gamma}(\lambda_\alpha, \leq)$ is a tree with the desired properties.

If $\gamma = \alpha + 1$, let $X = \lambda_\gamma - \lambda_\alpha$. Our plan is to add X^0 to $(\lambda_\alpha, \leq)$ as a λ_αth level in a special way, and then add T_X to this tree to get $(\lambda_\gamma, \leq)$.

Let $g: X^0 \to \lambda_\alpha$ be one-to-one and choose $\delta_0 < \delta_1 < \cdots$ having λ_α as a limit.

Case (1). Suppose that S_{λ_α} is a maximal antichain in $(\lambda_\alpha, \leq)$. For each $x \in X^0$ choose $x_0 < x_1 < \cdots$ in $(\lambda_\alpha, \leq)$ such that x_n belongs to the δ_nth level of $(\lambda_\alpha, \leq)$ and, for some $k \in \omega$, both $g(x)$ and some term of S_{λ_α} precede x_k in $(\lambda_\alpha, \leq)$.

Case (2). Otherwise do as in Case (1) without requiring that some term of S_{λ_α} precede an x_k.

Now let $(\lambda_\alpha \cup X^0, \leq)$ be the tree where $y \leq x$ if and only if (1) y and x belong to λ_α and $y \leq x$ in $(\lambda_\alpha, \leq)$ or (2) $y \in \lambda_\alpha$ and $x \in X^0$ and $y \leq x_n$ for some $n \in \omega$.

Define $(\lambda_\gamma, \leq) = (\lambda_\alpha \cup X^0, \leq) + (T_X)$.

The Souslin tree $(\omega_1, \leq) = \Sigma_{\alpha<\omega_1}(\lambda_\alpha, \leq)$. We only need to check that there are no uncountable antichains in $(\omega_1, \leq)$. So assume that S is a maximal antichain in $(\omega_1, \leq)$. Then $A = \{\lambda_\alpha \mid S \cap \lambda_\alpha$ is a maximal antichain in $(\lambda_\alpha, \leq)\}$ is a closed unbounded set in ω_1. Thus, by $\diamondsuit$, there is a $\lambda_\alpha \in A$ such that $S_{\lambda_\alpha} = S \cap \lambda_\alpha$. This is Case (1). Hence, if $x \in X^0$ which is the λ_αth level of $(\omega_1, \leq)$, x is preceded by a member of S. Therefore $S \subset (\lambda_\alpha, \leq)$ and S is countable.

(5) *Assume that* $(S, \leq)$ *is a special Aronszajn tree. We use* $\diamondsuit$, *as opposed to* $\diamondsuit$ (and a somewhat more sophisticated argument) *to prove that* S *with the tree topology is not normal.*

Since $(S, \leq)$ is special, S is the union of countably many antichains. Some antichain A in $(S, \leq)$ must intersect a stationary set of levels. Let T be the union of the levels of $(S, \leq)$ intersected by A. We prove that there are disjoint subsets of A which cannot be separated in T: thus S is not normal.

For $\alpha < \omega_1$, let T_α be the αth level of $(T, \leq)$. Choose $x_\alpha \in A \cap T_\alpha$ and let $X = \{x_\alpha\}_{\alpha \in \omega_1}$. For $\alpha < \beta$ in ω_1, let $P(\alpha, \beta) = \{t \in T \mid t \leq x_\beta$ and $\alpha <$ level of $t\}$.

Assume that $\{f_\alpha\}_{\alpha<\omega_1}$ is a family satisfying $\diamondsuit$.

For each $\alpha \in \omega_1$ we define a subset H_α of X by induction.

Define $H_0 = \emptyset$.

Suppose that $0 < \gamma < \omega_1$ and that H_α has been defined for all $\alpha < \gamma$.

If γ is a limit ordinal, define $H_\gamma = \bigcup_{\alpha<\gamma} H_\alpha$.

If $\gamma = \alpha + 1$, define $K_\alpha = \bigcup\{P(f_\alpha(\beta), \beta) \mid \beta \in \alpha$ and $x_\beta \in H\}$. If $x_\alpha \in \overline{K}_\alpha$, let $H_\gamma = H_\alpha$. If $x_\alpha \notin \overline{K}_\alpha$, let $H_\gamma = H_\alpha \cup \{x_\alpha\}$.

Define $H = \bigcup_{\alpha<\omega_1} H_\alpha$.

Since X is an antichain, if S is normal there are disjoint open sets U and V in T with $H \subset U$ and $(X - H) \subset V$. Thus, for each $0 < \alpha \in \omega_1$, there is $f(\alpha) < \alpha$ such that

$$H \subset \bigcup\{P(f(\alpha), \alpha) \mid x_\alpha \in H\} \subset U \quad \text{and} \quad (X - H) \subset \bigcup\{P(f(\alpha), \alpha) \mid x_\alpha \in X - H\} \subset V.$$

For $\alpha \in \omega_1$, define $F_\alpha = \bigcup\{P(f(\beta), \beta) \mid \beta \in \alpha\}$ and let $B = \{\alpha \in \omega_1 \mid x_\alpha \notin \overline{F}_\alpha\}$. If $\alpha \in B$ there is $f(\alpha) \leq g(\alpha) < \alpha$ such that $P(g(\alpha), \alpha) \cap F_\alpha = \emptyset$. Thus $\{P(g(\alpha), \alpha) \mid \alpha \in B\}$ are disjoint. But g is a pressing down function on B so B is not stationary; thus $\omega_1 - B$ contains a closed unbounded set.

Therefore, by $\diamondsuit$, there is an $\alpha \in (\omega_1 - B)$ such that $f \restriction \alpha = f_\alpha$. Since $x_\alpha \in \overline{F}_\alpha$, either $x_\alpha \in \overline{K}_\alpha$ or $x_\alpha \in \overline{G}_\alpha$ where $G_\alpha = \{P(f(\beta), \beta) \mid \beta \in \alpha$ and $x_\beta \in X - H\}$.

If $x_\alpha \in \overline{K}_\alpha$, then $x_\alpha \in X - H$. But $X - H \subset V$ and $K_\alpha \subset U$ which contradicts $U \cap V = \emptyset$. Similarly, if $x_\alpha \notin \overline{K}_\alpha$, then $x_\alpha \in H$ and $x_\alpha \in \overline{G}_\alpha$. But $H \subset U$ and $G_\alpha \subset V$. Thus S is not normal.

(6) Fleissner proves that, *if* $V = L$, *then every normal space of character* $\leq c$ *is collectionwise Hausdorff.* The technique is really quite similar to the proof of the preceding theorem. He proves by induction that the space is κ-collectionwise Hausdorff for all cardinals κ. One has a discrete closed set X of cardinality κ; functions $f: \kappa \to \omega_1$ correspond to the covering of X made by using the $f(\alpha)$th basic open set

for the αth point of X. One goes through X one point at a time constructing a set H which cannot be separated from $X - H$ by any function f. The basic tool is a $\diamondsuit$-like lemma for (regular) κ which is true in L; GCH is used at singular cardinals.

(7) In contrast with Fleissner's theorem which uses so much of the power of $V = L$, Juhász and Hajnal show *the existence of a first countable S-space using only the continuum hypothesis.* Recall from the previous chapter than an S-space is a regular, hereditarily separable, non-Lindelöf space. In (8) we give Ostaszewski's proof that $\diamondsuit$ implies the existence of a very powerful first countable S-space. But here we wish to give Juhász and Hajnal's beautiful combinatorial equivalence:

THEOREM. *There is a first countable S-space if and only if:*

(A) *There is a family* $\{A_{\alpha,n} \mid \alpha < \omega_1$ *and* $n \in \omega\}$ *of nonempty subsets of* ω *with the following properties:*

(a) $\alpha < \omega_1$ *implies* $A_{\alpha,0} \supset A_{\alpha,1} \supset \cdots$.

(b) $\beta < \alpha < \omega_1$ *implies there is* $n \in \omega$ *such that* $A_{\beta,0} \cap A_{\alpha,n} = \varnothing$.

(c) $\beta < \alpha < \omega_1$ *and* $n \in \omega$ *imply there is* $k \in \omega$ *such that either* $A_{\beta,k} \subset A_{\alpha,n}$ *or* $A_{\beta,k} \cap A_{\alpha,n} = \varnothing$.

(d) $M \subset \omega_1$ *is uncountable and* $n \in \omega$ *imply there are* $\beta < \alpha$ *in* M *and* $k \in \omega$ *such that* $A_{\beta,k} \subset A_{\alpha,n}$.

If (A) holds, a first countable S-space is obtained by topologizing ω_1 under the rule: $U \subset \omega_1$ is open and $\alpha \in U$ implies there is an $n \in \omega$ such that $U \supset \{\beta \in \omega_1 \mid A_{\beta,k} \subset A_{\alpha,n}$ for some $k \in \omega\}$.

On the other hand suppose that X is a first countable S-space. Since X is not Lindelöf, there is a subset $Y = \{x_\alpha\}_{\alpha<\omega_1}$ of X such that $\{x_\beta\}_{\beta\leq\alpha}$ is open in Y for all α. Since X is hereditarily separable, Y is separable and we assume that $\{x_n\}_{n\in\omega}$ is dense in Y. For each $\alpha < \omega_1$ choose a nested sequence $U_{\alpha,0} \supset U_{\alpha,1} \supset \cdots$ of open sets in Y, forming a basis for the topology of Y at x_α, such that $\overline{U}_{\alpha,0} \cap \{x_\beta\}_{\beta>\alpha} = \varnothing$. Define $A_{\alpha,n} = \{i \in \omega \mid x_i \in U_{\alpha,n}\}$; $\{A_{\alpha,n} \mid \alpha < \omega_1$ and $n \in \omega\}$ has the desired properties for (A).

Juhász and Hajnal use CH to index the real numbers (and subsets of the reals order isomorphic to the rationals) by countable ordinals; then by induction they choose sets satisfying (A). They also give a partial order form of (A).

(8) The existence of *Ostaszewski's space* is shocking partly because it destroys so many conjectures, but also because it is so easy to construct!

Assume that $\{S_\alpha\}_{\alpha<\omega_1}$ is a family satisfying $\clubsuit$.

Use CH to index the set of all countable infinite subsets of ω_1 as $\{X_\alpha\}_{\alpha<\omega_1}$ with $X_\alpha \subset \lambda_\alpha$ (where λ_α is the αth limit ordinal in ω_1).

For each $\beta < \omega_1$ and $n \in \omega$ we define a set $U_{\beta,n}$ by induction:

Define $U_{k,n} = \{k\}$ for all $k \in \omega$ and $n \in \omega$.

Now suppose that $0 < \gamma < \omega_1$ and that, for all $\alpha < \gamma$ and $\beta < \lambda_\alpha$, $U_{\beta,n}$ has been defined so that:

(1) $\{U_{\beta,n} \mid \beta < \lambda_\alpha$ and $n \in \omega\}$ is the basis for a Hausdorff topology T_α on λ_α.

(2) Each $U_{\beta,n}$ with $\beta < \lambda_\alpha$ and $n \in \omega$ is compact in T_α.

(3) $(\beta + 1) \supset U_{\beta,0} \supset U_{\beta,1} \supset \cdots$ for $\beta < \lambda_\alpha$ and $\{U_{\beta,n}\}_{n\in\omega}$ is a local basis for β in T_α.

If γ is a limit ordinal, then $\{U_{\beta,n} \mid \beta < \lambda_\gamma$ and $n \in \omega\}$ has been defined and (1), (2), and (3) are satisfied with γ in place of α.

So assume that γ is not a limit ordinal; say $\gamma = \alpha + 1$; we now define $U_{\beta,n}$ for all $\lambda_\alpha \leq \beta < \lambda_\gamma$.

Case (1). Suppose that X_α has no limit point in $(\lambda_\alpha, T_\alpha)$. Choose disjoint subsets $X = (x_0 < x_1 < \cdots)$ of X_α and $S = (s_0 < s_1 < \cdots)$ of S_α such that S is cofinal with λ_α. Observe that both X and S are closed in $(\lambda_\alpha, T_\alpha)$. For each $\beta \in X \cup S$ there is an n such that $U_{\beta,n} \cap (X \cup S) = \{\beta\}$; say $U_{\beta,n} = V_k$ if $\beta = x_k$ and $U_{\beta,n} = W_k$ if $\beta = s_k$. For each $k \in \omega$ define $V_k^* = V_k - \bigcup_{i<k}(V_i \cup W_i)$ and $W_k^* = W_k - \bigcup_{i<k}(V_i \cup W_i)$. Observe that x_k and s_k belong to $V_k^* \cup W_k^*$ which is compact and open in $(\lambda_\alpha, T_\alpha)$ and $\{V_k^* \cup W_k^*\}_{k\in\omega}$ are disjoint. Partition ω into infinitely many infinite disjoint subsets $N_0, N_1, \cdots$. Then, for each i and n in ω, define

$$U_{(\lambda_\alpha+i),n} = \{\lambda_\alpha + i\} \cup \bigcup\{V_k^* \cup W_k^* \mid k \in N_i \text{ and } k > n\}.$$

Since, for $i \neq j$, $U_{(\lambda_\alpha+i),0} \cap U_{(\lambda_\alpha+j),0} = \emptyset$, it is easy to check that (1), (2), and (3) hold for γ.

Case (2). If X_α has a limit point in (λ_α, T), make the exact construction described in Case (1) leaving out all mention of X_α (or X, x_k, V_k, or V_k^*). The whole purpose of the X_α's is to make sure that our space is countably compact: every countable infinite set needs a limit point but not necessarily more than one.

$\{U_{\beta,n} \mid \beta \in \omega_1$ and $n \in \omega\}$ is a basis for a topology T on ω_1; our space is (ω_1, T). Clearly (1), (2), and (3) are satisfied with $\alpha = \omega_1$ and T is countably compact.

Let us prove that open sets are either countable or co-countable. Suppose that V is open and that $\omega_1 - V$ is uncountable. By ♣ there is an $\alpha < \omega_1$ such that $S_\alpha \subset (\omega_1 - V)$. Since $\beta \geq \lambda_\alpha$ implies that $\beta \in \overline{S}_\alpha$, $V \subset \lambda_\alpha$ and thus V is countable. This fact shows that closed sets are G_δ sets and hereditary normality also easily follows. This same argument yields hereditary separability.

VII. Compactifications and βN

[73] (1) βN is compact, extremally disconnected and has 2^c points.

[73] (2) βN has exactly c easily described automorphisms.

[52] (3) Martin's axiom implies that $\beta N - N$ has a selective P-point.

[73] (4) CH implies that all P-points in $\beta N - N$ are of the same type.

[52] (5) (MA + $\neg$CH) implies there is a variety of P-points in $\beta N - N$.

[75] (6) If every dominating family has cardinality c, then there is a P-point in $\beta N - N$.

[76] (7) CH and $p \in \beta N - N$ imply that $\beta N - N - \{p\}$ is not normal.

[74] (8) $\beta N - N$ is not homogeneous; there are 2^c types of points in $\beta N - N$.

Compactness is one of the most important topological properties; important not only to topologists but to all mathematicians. Even when a space X is not compact, we may wish to study X as a subset of a compact space Y in which X is densely embedded; Y is called a compactification of X. If X is locally compact, its one point compactification is perhaps the most useful. If X is completely regular, X has a largest compactification known as its Čech compactification or βX [78], [21]; βX mirrors the continuous bounded real valued functions on X. Even a non-Hausdorff X has a Stone space [51]; X is embedded as a closed subset of its Stone space only if X is regular and extremally disconnected; but the Stone space still preserves many of the qualities of X as we have already seen. Real-compactifications, Wallman compactifications, the related topics seem endless.

For the purposes of this chapter we shall limit ourselves to looking at βN where $N = \omega$; even this subject is much too large and we will concentrate on P-point and automorphism problems in $\beta N - N$. But it is a good place to illustrate our theme. Compactification problems are topological: their motivation and applications are topological. But especially when looking at extremally disconnected compact spaces, there is an immediate translation of the problems into Boolean algebra terms: into purely set theoretic terms. The answers depend on the set theoretic assumptions; set theorists have often already looked at the problems. To attack such problems without intimate knowledge of modern set theory is foolhardy. To some extent this is true in all of abstract space theory, but in compactifications it is true is spades.

My favorite elementary introduction to βN and especially to P-point and automorphism problems is [73]. The bible in this area is, of course, [20]. Comfort and Negrepontis are just completing a new book [101] on the subject which will give the whole area a unification and set theoretic modernization which is badly needed. Actually, the field is moving so fast that they find it almost impossible to say, "now the book is complete"; but I presume we will see the book within a year and it will be the right place to look for a rich variety of references as well as an overview of the area.

(1) We define βN as the Stone space of ω. That is βN is the set of all ultrafilters on ω. If $M \subset \omega$, then we define $M' = \{x \in \beta N \mid M \in x\}$. βN is topologized by using $\{M' \mid M \subset \omega\}$ as a basis. If $\{M' \mid M \in \mathfrak{M}\}$ is a family of basic open sets, then $\{\omega - M \mid M \in \mathfrak{M}\}$ is a filter if and only if $\{M' \mid M \in \mathfrak{M}\}$ is not a cover of βN; thus βN is compact. Also $\overline{\bigcup\{M' \mid M \in \mathfrak{M}\}}$ and $(\omega - \bigcup\{M \mid M \in \mathfrak{M}\})'$ are disjoint open sets covering βN; so βN is extremally disconnected.

To prove that *βN has 2^c points* find a family $\{M_\alpha\}_{\alpha<c}$ of subsets of ω such that, if $(\alpha_1, \alpha_2, \cdots, \alpha_n)$ and $(\beta_1, \beta_2, \cdots, \beta_k)$ are disjoint finite subsets of c, then

$$M_{\alpha_1} \cap M_{\alpha_2} \cap \cdots \cap M_{\alpha_n} \cap (\omega - M_{\beta_1}) \cap \cdots \cap (\omega - M_{\beta_k}) \neq \emptyset.$$

Such a family (called an independent family) can be found by a simple Cantor tree construction. If S is a subset of c, let x_S be an ultrafilter on ω containing $\{M_\alpha\}_{\alpha \in S}$ and $\{\omega - M_\alpha\}_{\alpha \in c - S}$. Clearly all of the x_S are different; hence βN has 2^c points. The reference I give for this is Walter Rudin's paper [73] which is well written and easily available; however, this theorem was known to Hausdorff around 1915.

(2) Rudin also points out that the homeomorphisms of βN onto βN are all quite simple:

If $n \in \omega$ there is an ultrafilter n^* consisting exactly of those subsets of ω containing n; $\{n\}'$ is an open set in βN whose only element is n^*. We identify n with n^*; thus $N = \{n^* \mid n \in \omega\}$ is an open, countable, discrete subset of βN whose closure is βN; N is a homeomorphic copy of ω embedded in βN as a dense subset. Any homeomorphism of βN onto βN must take N onto N. Therefore, *each permutation of N (or ω) induces a homeomorphism of βN onto βN and vise versa.*

We say that two points p and q of a topological space X are *of the same* type in X provided there is a homeomorphism from X onto X taking p into q. One consequence of the above theorem is that there are 2^c types of points in βN.

βN is really very rigid; however $\beta N - N$ is much more complex. We will use the convenient standard notation of N^* for $\beta N - N$. If $M \subset \omega$, let $M^* = \{x \in N^* \mid M \in x\}$. Clearly $\{M^* \mid M \subset \omega\}$ is a basis for the topology of N^*; and if $M \subset \omega$ and $L \subset \omega$ then $M^* \subset L^*$ if and only if $M - L$ is finite.

A P-point in a topological space X is a point x such that, for all countable families $\{U_i\}_{i \in \omega}$ of open sets with $x \in \bigcap_{i \in \omega} U_i$ there is an open set U with $x \in U \subset \bigcap_{i \in \omega} U_i$. No separable space, for instance βN, can have a P-point. If $x \in N^*$, x is a P-point provided, for every countable subset $\{M_i\}_{i \in \omega}$ of x, there is an $M \in x$ such that $|M - M_i|$ is finite for all i. Let $L_i = \bigcap_{j \le i} M_j - \bigcup_{j > i} M_j$; we say that x is a *selective* P-point if M can always be chosen with $|M \cap L_i| \le 1$. Rudin [73] proves that CH implies there is a P-point in N^*; Booth [52] proves that Martin's axiom implies there is a selective P-point in N^*. Kunen [77] has shown that it is consistent that there be no selective P-points in N^*. It is conjectured but unproven that there are P-points in N^* without any set theoretic assumptions.

A filter F is said to *free* provided the intersection of any finite subset of F is infinite. The ultrafilters in N^* are precisely the free ultrafilters in βN.

(3) By Booth's theorem, (1) of Chapter IV, Martin's axiom implies that if F is a free filter on ω of cardinality less than c, then there is an infinite subset A of ω such that $A - B$ is finite for all $B \in F$.

Using Martin's axiom let us construct a selective P-point in N^* by induction on c. We can index the set of all partitionings of ω into countably many disjoint sets by $\{L_\alpha\}_{\alpha<c}$ where $L_\alpha = \{L_{\alpha,n}\}_{n\in\omega}, \cdots$. Define $M_{\alpha,n} = \bigcup_{j\geq n} L_{\alpha,j}$. For each $\alpha < c$ we choose F_α as follows.

Define $F_0 = \{N\}$.

Suppose $\gamma < c$ and that, for all $\alpha < \gamma$, a free filter F_α of cardinality less than c has been chosen with $F_\beta \subset F_\alpha$ for all $\beta < \alpha$.

If γ is a limit ordinal, define $F_\gamma = \bigcup_{\alpha<\gamma} F_\alpha$.

If $\gamma = \alpha + 1$ we have two cases.

Case (1). Suppose that $F_\alpha \cup \{M_{\alpha,n}\}_{n\in\omega}$ is a free filter. By Martin's axiom and Booth's theorem there is an infinite subset A of ω such that $A - B$ is finite for all $B \in F_\alpha \cup \{M_{\alpha,n}\}_{n\in\omega}$. For each $n \in \omega$ for which it is possible, choose $x_n \in A \cap L_{\alpha,n}$. Then $X = \{x_n\}_{n\in\omega}$ is infinite and $|X \cap L_{\alpha,n}| \leq 1$. Define $F_\gamma = F_\alpha \cup \{X\}$.

Case (2). If $F_\alpha \cup \{M_{\alpha,n}\}_{n\in\omega}$ is not a free filter, define $F_\gamma = F_\alpha$.

Clearly any ultrafilter containing $\bigcup_{\alpha<\omega_1} F_\alpha$ is a selective P-point in N^*.

(4) The following theorem aptly illustrates why it is so useful to assume CH when working with N^*.

Suppose that p and q are P-points in N^* and let $\mathcal{B}$ be the set of all basic open (compact) sets in N^*. If CH, then we can index $\mathcal{B} = \{U_\alpha\}_{\alpha<\omega_1}$. We describe a permutation f of $\mathcal{B}$ which induces a homeomorphism of N^* onto N^* taking p to q.

Let us call a subset of $\mathcal{B}$ a *ring* if it is closed under finite unions, finite intersections, and complements.

We define f by induction; at stage α the domain of f is $\mathcal{R}_\alpha$, the smallest ring containing a subset $\mathcal{B}_\alpha$ of B. We assume that

(1) $\mathcal{R}_\alpha$ is countable and $f \restriction \mathcal{R}_\alpha$ preserves finite unions, finite intersections, and complements.

(2) $p \in X \in \mathcal{R}_\alpha$ if and only if $q \in f(X)$.

(3) For $\beta < \alpha$, $\mathcal{R}_\beta \subset \mathcal{R}_\alpha$.

Define $\mathcal{B}_0 = \{N^*\}$ and $f(N^*) = N^*$. At limit ordinals γ define $\mathcal{B}_\gamma = \bigcup_{\alpha<\gamma} \mathcal{B}_\alpha$. So suppose f has been defined for $\mathcal{R}_\alpha$ satisfying our hypotheses. Let Z be the one of U_α and $N^* - U_\alpha$ containing p.

Choose U and V in $\mathcal{B}$ with $p \in U$, $q \in V$ and U and V properly contained in $Z \cap \bigcap\{X \in \mathcal{R}_\alpha \mid p \in X\}$ and $\bigcap\{f(X) \mid q \in f(X)$ and $X \in \mathcal{R}_\alpha\}$, respectively. Define $f(U) = V$ and let $\mathcal{R}$ be the smallest ring containing $\mathcal{R}_\alpha \cup \{U\}$.

For each $X \in \mathcal{R}$ such that $X \cap Z \neq \emptyset$ choose a P-point $z_X \in X \cap Z$; choose $W_X \in \mathcal{B}$ such that $W_X \subset \bigcap\{f(Y) \mid Y \in \mathcal{R}$ and $z_X \in Y\}$. Define $f(Z) = \bigcup\{f(X) \mid X \in \mathcal{R}$ and $X \subset Z\} \cup \bigcup\{W_X \mid X \cap Z \neq \emptyset\}$.

Symmetrically add Z to the range of f and then define $\mathcal{B}_\gamma$ to be the resulting domain of f; then extend f to R_γ. When f is defined for all $\alpha < \omega_1$, f clearly induces a homeomorphism of N^* which takes p onto q. This technique is often useful [82].

(5) *If we assume* (MA + $\neg$CH) *there are P-points of different types in* N^*. To see this it may help to state Booth's theorem ((1) of Chapter IV) in a more topological form:

BOOTH'S THEOREM. *Martin's axiom implies that if F is a filter of basic open sets in N^* of cardinality less than c, then there is a family $\{U_\alpha\}_{\alpha<c}$ of basic open sets in N^* such that for all $\alpha < \beta < c$, $U_\beta \subset U_\alpha \subset \bigcap F$.*

Assume (MA + $\neg$CH). There are obviously selective P-points in N^* which are the intersection of c well ordered by inclusion basic open sets. The technique of (4) shows that all such P-points are of the same type in N^*. Choose a strictly decreasing family $\{V_\alpha\}_{\alpha<\omega_1}$ of basic open sets in N^*. There is clearly a P-point in $\bigcap_{\alpha<\omega_1} V_\alpha$ which is in the closure of $\{N^* - V_\alpha\}_{\alpha<\omega_1}$. Such a P-point is not of the same type as those discussed earlier. The variety of P-points one can build obviously depends on the size of the continuum.

One should observe that CH implies that there are P-points in N^* which are not selective even though they are all of the same type in N^*.

(6) Let F be the set of all functions from ω into ω. If f and g belong to F, define $f \leq g$ provided $\{n \in \omega \mid f(n) > g(n)\}$ is finite. A subset G of F is *dominant* if, for every $f \in F$, there is a $g \in G$ such that $f \leq g$. A *scale* is a well ordered by $<$, increasing, dominating family. It should come as no surprise that there are relations between the existence of scales and dominating families of various cardinalities and the existence of p-points in N^*. A source of theorems in this area is [104].

Booth's theorem says that with Martin's axiom there is a scale of cardinality c and there is no dominating family of cardinality less than c.

Under other hypotheses it is consistent to assume:

(1) There is no scale but there is a dominating family of cardinality less than c.

or (2) There is a scale of cardinality less than c.

or (3) There is a dominating family of cardinality α but there is no scale of cardinality α although there is a scale.

KETONEN'S THEOREM. (H) *If and only if* (K).

(H) *No family of cardinality less than c dominates.*

(K) *Every filter on ω of cardinality less than c can be extended to a P-point.*

Assume (H) and that $\mathcal{F}$ is a filter on ω of cardinality less than c. We show that, if $\{X_n\}_{n\in\omega}$ is a subset of $\mathcal{F}$ with $X_0 \supset X_1 \supset \cdots$, then there is an infinite subset X of ω such that $\mathcal{F} \cup \{X\}$ is a filter and $|X - X_n|$ is finite for all $n \in \omega$. By induction one can then prove (K).

Without loss of generality we assume that all intersections of finite subsets of $\mathfrak{F}$ are infinite. For each $Y \in \mathfrak{F}$, we define $f_Y\colon \omega \to \omega$ by $f_Y(n)$ is the smallest integer in $(X_n \cap Y)$. By (H) there is an f such that $f \not\leq f_Y$ for any $Y \in \mathfrak{F}$. Define

$$X = \bigcup_{n \in \omega} (X_n \cap \{j \in \omega \mid j \leq f(n)\});$$

this X has the desired properties.

Now assume (K) and that G is a subset of F of cardinality less than c. Arbitrarily partition ω into disjoint infinite subsets, say $J_0, J_1, \cdots$. For each $g \in G$, define

$$X_g = \bigcup_{n \in \omega} \{j \in J_n \mid g(n) \leq j\}.$$

By (K), there is an $X \subset \omega$ such that $X \cap J_n$ is finite for all $n \in \omega$, and $X \cup \{X_g\}_{g \in G}$ is a filter. Choose $f(n) \in X \cap J_n$ if $X \cap J_n \neq \emptyset$; otherwise define $f(n) = 0$. Then $f \not\leq g$ for any $g \in G$. Thus G does not dominate and (H) holds.

(7) We prove the following lemmas:

(a) WARREN'S LEMMA [75]. *If D is an uncountable discrete set and $E = \{p \in \beta D \mid p \in \overline{X}$ for some countable $X \subset D\}$, then E is not normal.*

(b) COMFORT AND NEGREPONTIS' LEMMA [82]. *CH and $p \in N^*$ imply that there is a closed subset K of N^* such that $K - \{p\}$ is homeomorphic to E.*

An immediate consequence of (a) and (b) is that *CH implies that $N^* - \{p\}$ is not normal for all $p \in N^*$.*

PROOF OF (b). Suppose that $p \in N^*$ and that $\{U_\alpha\}_{\alpha<\omega_1}$ is a basis for the topology of N^* at p. Let $\mathcal{C}$ be the set of all countable sets C of P-points with $p \in \overline{C}$. Observe that if B and C belong to $\mathcal{C}$, then $p \notin \overline{B - C}$. If $\mathcal{C} \neq \emptyset$, select $C \in \mathcal{C}$; if $\mathcal{C} = \emptyset$, let $C = \{p\}$. By induction, for each $\alpha < \omega_1$ choose a P-point $p_\alpha \in \bigcap_{\beta<\alpha} U_\beta - \overline{C}$. Then $D = \{p_\alpha\}_{\alpha<\omega_1}$ is discrete in N^* and $K = \overline{D}$ has the desired properties.

PROOF OF (a). Let $D = (\omega_1 \times \omega)$ with the discrete topology and let $\{\lambda_\alpha\}_{\alpha<\omega_1}$ be the order preserving indexing of the limit ordinals in ω_1. By induction, for each $\alpha < \omega_1$, choose disjoint subsets X_α, Y_α of $(\lambda_\alpha \times \omega)$ such that:

(1) For all $n \in \omega$ there is a $\beta < \lambda_\alpha$ such that $\{(\gamma, n) \mid \beta < \gamma < \lambda_\alpha\} \subset Y_\alpha$.

(2) For all $\beta < \lambda_\alpha$ there is an $n \in \omega$ such that $\{(\beta, k) \mid k > n\} \subset X_\alpha$.

(3) $\alpha < \beta$ implies that $|X_\alpha - X_\beta|$ and $|Y_\alpha - Y_\beta|$ are finite.

Let $X = \bigcup_{\alpha<\omega_1}$ (closure of X_α in βD) and $Y = \bigcup_{\alpha<\omega_1}$ (closure of Y_α in βD). Then $(X - D)$ and $(Y - D)$ are closed and disjoint in E but cannot be separated by disjoint open sets. Hence E is not normal.

(8) It would be nice to have some simple, useful way of classifying the homeomorphisms of N^* onto N^*. This goal is still far away. One attempt to solve this problem was the study of a number of natural partial orders on N^* [79]; none of them distinguishes between types of points in N^* but their study has become an end in itself for they do yield information. A. Blass [80] has proved that most of the theorems

about these partial orders which were originally proved using CH hold even assuming Martin's axiom.

Let S be the set of all simple countable discrete sequences of points in βN. If $p \in \beta N$ and $X = (x_0, x_1, \cdots) \in S$, define $p(X)$ to be the image of p under the natural homeomorphism of βN onto $\overline{X}$ which takes n into x_n. If p and q belong to βN, define $p \leq q$ provided there is an $X \in S$ such that $p(X) = q$. Then $\leq$ is a partial order on βN where $p = q$ means that p and q are of the same type in βN. Let us make three observations which were first made by Frolik [74]: suppose $p \in \beta N$.

(a) $|\{q \in \beta N \mid q \leq p\}| \leq c$.

(b) $|\{q \in \beta N \mid p \leq q\}| = 2^c$.

(c) $\{q \in \beta N \mid q \leq p\}$ *is totally ordered by* $\leq$.

PROOF OF (a). If $q \leq p$ there is an $X \in S$ such that $q(X) = p$. Since $X = x_{0,q}, x_{1,q}, \cdots$ is discrete, there are disjoint subsets $M_{0,q}, M_{1,q}, \cdots$ of ω such that $M_{n,q} \in x_{n,q}$. Suppose that $q \leq p$ and $r \leq p$ and that $M_{n,q} = M_{n,r}$ for all $n \in \omega$. For each $n \in \omega$ let Q_n and R_n be disjoint subsets of $M_{n,q}$ with $Q_n \in x_{n,q}$ and $R_n \in x_{n,r}$. Since either $\bigcup_{n\in\omega} Q_n \notin p$ or $\bigcup_{n\in\omega} R_n \notin p$, it is impossible that both $q \leq p$ and $r \leq p$. Hence the number of $q \leq p$ is at most the number of partitionings of ω into disjoint sets which is c.

PROOF OF (b). Arbitrarily partition ω into infinitely many disjoint infinite sets $M_0, M_1, \cdots$. Arbitrarily index $\{x \in \beta N \mid M_n \in x\} = \{x_{\alpha,n}\}_{\alpha<2^c}$. For $\alpha < 2^c$, let $X_\alpha = x_{\alpha,0}, x_{\alpha,1}, \cdots$. Then $\{p(X_\alpha)\}_{\alpha<2^c}$ is a family of 2^c points in βN greater than p.

PROOF OF (c). Suppose that there exist X and Y in S and q and r in βN such that $q(X) = p$ and $r(Y) = p$. Clearly p belongs to the closure of one of the following sets: (1) $X \cap Y$, (2) $X - \overline{Y}$, (3) $X \cap (\overline{Y} - Y)$, (4) $Y - \overline{X}$ or (5) $Y \cap (\overline{X} - X)$. In case (1), q and r are of the same type in βN. We cannot have both cases (2) and (4). If not cases (2) or (1), then case (3) and $r < q$. If not cases (4) or (1), then case (5) and $q < r$. Hence $q \leq p$ and $r \leq p$ implies either $q \leq r$ or $r \leq q$.

These three facts give us a rather clear picture of βN as reflected in Frolik's ordering: (c) tells us that we have a tree-like structure; (a) tells us that it is rather small looking back; and (b) tells us that it is very thick looking up. The fixed ultrafilters are at the base. If two points are of the same type in N^*, then they have the same set of predecessors in the tree. But, unfortunately there are three types of candidates for minimal elements in N^*:

(1) P-points,

(2) points which are limit points of some countable subset of N^* but are not limit points of any countable discrete subset,

(3) points which are neither P-points nor limit points of any countable subset of N^*.

Kunen has proved [75] that all three types of points exist if we assume CH.

Frolik uses (a) to get an immediate proof that N^* *is not homogeneous; in fact,* N^* *has* 2^c *types of points*. Choose an arbitrary $X \in S$ with $X \subset N^*$. If $x \in \overline{X} - X$ there is a unique $p \in N^*$ such that $p(X) = x$. Observe that $p < p(X) = x$. So if $y \in \overline{X} - X$ and x and y are of the same type in N^*, $p < y$. Thus, by (a), at most c points of $\overline{X}$ are of the same type in N^*. Since $\overline{X}$ has 2^c points, there are 2^c types of points in N^*.

Let us close with a few comments.

You may recall Ramsey's theorem $\omega \to (\omega)^r_n$. An ultrafilter on ω is called a *Ramsey ultrafilter* if, for every function f: $[\omega]^r \to n$, there is a homogeneous set with respect to this fucntion in the ultrafilter. An ultrafilter is Ramsey if and only if it is selective.

In the "random real" model of ZFC + $\neg$CH there is no selective ultrafilter; it is not known whether there is a P-point or not.

In Cohen's original model [19] which has $c = \aleph_2$, there are selective P-points but no "$\aleph_2$- P-points". That is, every point p of N^* is a limit point of the union of some cardinality $\aleph_1$ set of closed sets to which p does not belong even though $\aleph_1 < c$.

VIII. Metrization and Moore Spaces

[9] (1) Every collectionwise normal Moore space is metrizable.

[59] (2) Every separable, metacompact normal Moore space is metrizable.

[72] (3) Every locally compact, locally connected, normal Moore space is metrizable.

[40], [72] (4) Assuming Lusin's hypothesis, there is no separable normal nonmetrizable Moore space; and assuming $V = L$ there is no locally compact normal nonmetrizable Moore space.

[72] (5) $V = L$ implies every normal, first countable space in which every subset is on F_σ is σ-discrete.

[54], [55], [39] (6) (MA + $\neg$CH) implies a variety of normal nonmetrizable Moore spaces.

[9] (7) Bing's G.

[57] (8) Fleissner's George.

Metrization is the heart and soul of general topology. We now juggle an incredible collection of different classes of spaces which are almost, but not quite, metrizable. Some are obviously more important than others and it would probably be healthy for the area if we could erase all but a few of them. However, the delicate differences allow for more accurate theorems. Perhaps Hodel or Heath or someone with some of their insight ([83], [103]) will write a book for us unifying this complex area. No attempt will be made here to look at the broad picture. Instead this chapter will be devoted strictly to the normal Moore space conjecture [40]. This hard, basic, metrization problem has been yielding to set theoretic attack; and to attack it without some knowledge of set theory is madness; we again illustrate our point.

(1) We begin with the basic tool. *Suppose that X is a normal, collectionwise normal space with a development $\mathcal{G}_0, \mathcal{G}_1, \cdots$. We prove Bing's well-known theorem* [9] *that X is metrizable.*

Let $\mathcal{G}_n = \{U_{\alpha n}\}_{\alpha<\kappa}$. For each n and k in ω and $\alpha < \kappa$, define $C_{\alpha,n,k}$ to be the set of all points p of X such that:

(a) α is the smallest ordinal with $p \in U_{\alpha n}$ (n fixed).

(b) $\mathcal{G}^*_k(p) = \bigcup\{V \in \mathcal{G}_k \mid p \in V\} \subset U_{\alpha n}$.

For n and k fixed, $\{C_{\alpha,n,k}\}_{\alpha<\kappa}$ is a closed discrete family of closed sets. Since X is collectionwise normal, there is a family $\{V_{\alpha,n,k}\}_{\alpha<\kappa}$ of disjoint open sets with $C_{\alpha,n,k} \subset V_{\alpha,n,k} \subset U_{\alpha n}$. Then $\{V_{\alpha,n,k} \mid \alpha < \kappa,\ n \in \omega,\ k \in \omega\}$ is a σ-discrete basis for X. Hence X is metrizable.

(2) It is easy to prove Traylor's theorem that *a metacompact, separable, normal Moore space X is metrizable* since a metacompact separable space is Lindelöf and a Lindelöf-Moore space is metrizable.

(3) Somewhat more difficult is the new theorem of Reed and Zenor. *Assume that X is a locally compact, locally connected normal Moore space. We prove that every component Y of X is metrizable.* Since the components of a locally connected space are open and disjoint, we will thus prove that X is metrizable.

(a) $|Y| \leq c$. Each point p of Y belongs to a connected open subset U_p of Y with $|U_p| \leq c$. Choose $p_0 \in Y$ and by induction define U_α for each $\alpha < \omega_1$ as follows. Define $U_0 = U_{p_0}$ and, for $\alpha > 0$, define $U_\alpha = \bigcup\{U_p \mid p \in \overline{\bigcup_{\beta<\alpha} U_\beta}\}$. Since Y is first countable it is easy to prove by induction that $|U_\alpha| \leq c$ for all α. Since Y is connected and first countable, $\bigcup_{\alpha<\omega_1} U_\alpha = Y$; hence $|Y| \leq c$.

(b) *There is a countable separating open cover of Y.* Let $\mathcal{G}_0, \mathcal{G}_1, \cdots$ be a development for Y. By (a), there is a subset $\{U_{\alpha,n}\}_{\alpha<c}$ of $\mathcal{G}_n$ covering Y. For n and k in ω and $\alpha < c$, define $V_{\alpha,n,k} = U_{\alpha,n} - \bigcup_{\beta<\alpha}\{p \in Y \mid \mathcal{G}_k^*(p) \subset U_\beta\}$. Observe that $\{V_{\alpha,n,k}\}_{\alpha<c}$ is an open cover of Y. Define $f\colon c \to$ irrationals. Then, for each rational r, let $W_{r,n,k} = \bigcup\{V_{\alpha,n,r} \mid f(\alpha) < r\}$ and $Z_{r,n,k} = \{V_{\alpha,n,r} \mid f(\alpha) > r\}$. Then the set $\mathcal{G}$ of of all $W_{r,n,k}$ and $Z_{r,n,k}$ is a countable open cover of Y and, if $x \neq z$ in Y, there are W and Z in $\mathcal{G}$ such that $W \cup Z$ covers Y and $x \in W - \overline{Z}$ and $z \in Z - \overline{W}$.

(c) *Suppose that* $\{K_\alpha\}_{\alpha\in A}$ *is a closed discrete family of closed sets in Y. We prove that Y is metric* using (1) by finding disjoint open sets $\{U_\alpha\}_{\alpha\in A}$ with $K_\alpha \subset U_\alpha$.

Assume that $\mathcal{G} = \{W_n\}_{n\in\omega}$ is the separating cover of Y described in (b). For each $\alpha \in A$ and $n \in \omega$ and $x \in K_\alpha$, choose a connected open neighborhood $N_n(x)$ such that $\overline{N_n(x)}$ is compact, $N_n(x)$ is contained in a term of $\mathcal{G}_n$, $\overline{N_n(x)} \cap K_\beta = \emptyset$ for $\beta \neq \alpha$, $N_n(x) \subset \bigcap\{W_i \mid i \leq n \text{ and } x \in W_i\}$, and $N_0(x) \supset N_1(x) \supset \cdots$.

CLAIM. *If* $x \in K_\alpha$ *there is an* $n \in \omega$ *such that for all* $y \in \bigcup_{\beta\neq\alpha} K_\beta$, $N_n(x) \cap N_n(y) = \emptyset$. If such an n exists for each x we call it n_x and define $U_\alpha = \bigcup_{x\in K_\alpha} N_{n_x}(x)$. Clearly $\{U_\alpha\}_{\alpha\in A}$ are disjoint and $K_\alpha \subset U_\alpha$ so we have proved that Y is collectionwise normal and hence metric.

To prove our claim, assume that, on the contrary, for each $n \in \omega$ there is a $y_n \in \bigcup_{\beta\neq\alpha} K_\beta$ such that $N_n(y_n) \cap N_n(x) \neq \emptyset$. Since $N_n(y_n)$ is connected, for all $n > 1$ there is a $z_n \in N_n(y_n) \cap (N_1(x) - N_2(x))$. Since $\overline{N_1(x)}$ is compact, there is a point z every neighborhood of which contains z_n for infinitely many n. There are i and j in ω such that $W_i \cup W_j$ covers Y and $x \in W_i - \overline{W}_j$ and $z \in W_j - \overline{W}_i$. There is an $m > i + j$ such that $\mathcal{G}_m^*(x) \subset W_i - W_j$ and there is an $n > m$ with $N_n(y_n) \cap (W_j - \overline{W}_i) \neq \emptyset$. Therefore $N_n(y_n) \subset W_j$ and $N_n(x) \subset (W_i - W_j)$ so $N_n(y_n) \cap N_n(x) = \emptyset$ contrary to our hypothesis.

Our proof that Y and X are metric is complete.

(4) The above absolute results are very satisfying and we have other related positive consistency results. Jones' nice theorem, that $2^{\aleph_0} < 2^{\aleph_1}$ implies that separable normal Moore spaces are metrizable [40], was discussed earlier and makes the *metacompact* in Traylor's theorem seem unnecessary.

By far the most powerful theorem in the area we also discussed earlier: Fleissner's theorem [72] that $V = L$ implies that every normal space of character at most c is collectionwise Hausdorff. Fleissner says that collectionwise Hausdorff is just a consolation prize for people who cannot get collectionwise normality. There is some truth in this for many theorems have collectionwise normality in their hypotheses and few use collectionwise Hausdorff.

One consequence of Fleissner's theorem, however, makes the locally connected in Reed and Zenor's theorem look unnecessary. For $V = L$ *implies that all locally compact normal Moore spaces are metric.* To see this suppose that $\mathcal{G}_0, \mathcal{G}_1, \cdots$ is a development for a normal Moore space X and that all terms of $\bigcup_{n\in\omega}\mathcal{G}_n$ have compact closures. From the proof of Bing's theorem (1) we see that X is metrizable unless there is an n and a family $\{C_\alpha\}_{\alpha<\kappa}$ of closed sets with each C_α contained in a member of $\mathcal{G}_n$, such that $\{C_\alpha\}_{\alpha<\kappa}$ cannot be separated by disjoint open sets. In this case each C_α is compact. Let Y be the quotient space obtained by collapsing each C_α to a point c_α. By Fleissner's theorem, since the compactness of the C_α's keeps Y first countable, there are disjoint open sets separating the c_α's in Y. The same sets with C_α in place of c_α separate the C_α's in X.

(5) Another consequence of Fleissner's theorem observed by Reed is that $V = L$ *implies that every normal, first countable space* X, *every subset of which is an* F_σ, *is* σ*-discrete.*

For each $x \in X$ choose a local basis $\{U_n(x)\}_{n\in\omega}$ for the topology of X at x.

Let $Y = X \times (\omega + 1)$ which we topologize by declaring a set V to be open if and only if, for all $(x, \omega) \in V$ there is a $k \in \omega$ and, for each $i > k$ in ω, an $n_i \in \omega$ such that $\bigcup_{i>k}(U_{n_i} \times \{i\}) \subset V$.

Clearly Y has character c (and is Hausdorff). We prove Y is normal. Suppose H and K are disjoint closed subsets of Y; without loss of generality $(H \cup K) = (X \times \{\omega\})$. Let $H' = \{x \in X \mid (x, \omega) \in H\}$ and $K' = \{x \in X \mid (x, \omega) \in K\}$. Since every subset of X is an F_σ, $H' = \bigcup_{n\in\omega} H_n$ and $K' = \bigcup_{n\in\omega} K_n$ where H_n and K_n are closed in the topology of X and $H_0 \subset H_1 \subset \cdots$ and $K_0 \subset K_1 \subset \cdots$. Since H_n and K_n are closed and disjoint and X is normal, there are disjoint open sets V_n and W_n of X with $H_n \subset V_n$ and $K_n \subset W_n$. Define $V = H \cup \bigcup_{n\in\omega}(V_n \times \{n\})$ and $W = K \cup \bigcup_{n\in\omega}(W_n \times \{n\})$. Then U and V are open in Y, disjoint, and $H \subset V$ and $K \subset W$. Thus Y is normal and, by Fleissner's theorem, collectionwise Hausdorff.

Therefore, for each $x \in X$, there is an open set T_x in Y such that $(x, \omega) \in T_x$ and $\{T_x\}_{x\in X}$ are disjoint. For each $x \in X$ there is an i_x and an n_x in ω with $(U_{n_x}(x) \times \{x\}) \subset T_x$. For each $i \in \omega$ define $D_i = \{x \in X \mid i_x = i\}$. If x and y are in D_i, then $U_{n_x} \cap U_{n_y} = \emptyset$ since $T_x \cap T_y = \emptyset$. Thus each D_i is discrete, and $X = \bigcup_{i\in\omega} D_i$, so X is σ-discrete.

(6) In Chapter IV we showed that (MA + $\neg$CH) implies there are normal nonmetrizable Moore spaces of the following varieties:

(a) metacompact (Pixley-Roy)

(b) separable, locally compact (Cantor tree)

(c) separable, locally connected (connected Cantor tree)

(d) nonseparable, locally compact (Aronszajn tree)

(e) nonseparable, locally connected (connected Aronszajn tree).

I conjecture that there is a "real" example of a normal nonmetrizable Moore space. We now know that it cannot be separable or locally compact and it must be collectionwise Hausdorff. The only real steps in this direction are the following two examples separated by more than twenty years.

(7) BING'S EXAMPLE G. Let P be the set of all subsets of ω_1 and G be the set of all functions from P into the pair $(0, 1)$. For each $\alpha \in \omega_1$, let f_α be the characteristic function of α: that is, for $A \subset \omega_1$, $f_\alpha(A) = 1$ if and only if $\alpha \in A$. For $A \in P$ and $e = 0$ and 1, define $U_{A,e} = \{g \in G \mid g(A) = e\}$. We use $\{U_{A,e} \mid A \in P$ and $e = 0$ or $1\} \cup \{\{f\} \mid f \in G - \{f_\alpha\}_{\alpha<\omega_1}\}$ as a subbasis for the topology of G. It is easy to prove that *this space is normal but not collectionwise Hausdorff* because $\{f_\alpha\}_{\alpha<\omega_1}$ is a closed discrete subset of G which cannot be separated by disjoint open sets. Since G has character 2^{ω_1} which is 2^c if one assumes $V = L$, G does not violate Fleissner's theorem.

What we need then is a space of smaller character which is collectionwise Hausdorff as well as normal, but still not collectionwise normal. George has these properties, but has a long way to go to become a Moore space; its character is c.

(8) FLEISSNER'S GEORGE. Let D be ω_1 with the discrete topology and define X to be the subspace of $\omega_1 \times D$ consisting of those (β, α) with $\alpha < \beta$.

For each $\beta < \omega_1$, define $P_\beta = \{A \subset (\{\beta\} \times D) \mid A \subset X\}$ and $G_\beta = \{g\colon P_\beta \to (0, 1)\}$. For each $\alpha < \beta$, let $g_{\beta,\alpha}$ be the term of G_β such that, for each $A \in P_\beta$, $g_{\beta,\alpha}(A) = 1$ if and only if $(\beta, \alpha) \in A$.

Let $F = \bigcup_{\beta<\omega_1}(G_\beta - \{g_{\beta,\alpha} \mid \alpha \in \beta\})$ and, if V is clopen in X and $e = 0$ or 1, let $U_{V,e} = \bigcup_{\beta<\omega_1}\{g \in G_\beta \mid g(\beta,\alpha) = e$ for some $(\beta, \alpha) \in V\}$.

Define George $= \bigcup_{\beta<\omega_1} G_\beta$. Subbasic elements for the topology of George are either:

(a) $\{f\}$ for some $f \in F$,

(b) $\bigcup_{\beta \in I} G_\beta$ for some clopen I in ω_1, or

(c) $U_{V,e}$ for some clopen V in X and $e = 0$ or 1.

Let us check that George has the desired properties.

(i) *Hausdorff*. Suppose that $x \in G_\beta$ and $y \in G_\alpha$. If $\alpha = \beta$ and $x \neq y$, there is an $A \in P_\beta$ with $x(A) \neq y(A)$. If $B = \{(\gamma, \delta) \in X \mid (\alpha, \delta) \in A\}$, then $U_{B,0}$ and $U_{B,1}$ separate x and y. If $\alpha < \beta$, then $\bigcup_{\gamma\leq\alpha} G_\gamma$ and $\bigcup_{\gamma>\alpha} G_\gamma$ separate x and y.

(ii) *Normal*. Suppose that H and K are closed and disjoint in George. Let $H' = \{(\beta, \alpha) \in X \mid g_{\beta,\alpha} \in H\}$ and $K' = \{(\beta, \alpha) \in X \mid g_{\beta,\alpha} \in K\}$. Since H' and K' are closed in X, there is a clopen subset V of X with $H' \subset V$ and $K' \subset X - V$. Then $(U_{V,1} \cup (H \cap F)) - (K \cap F)$ and $(U_{V,0} \cup (K \cap F)) - (H \cap F)$ separate H and K.

(iii) *Collectionwise Hausdorff*. Suppose that Z is a closed discrete subset of George. Assume that $Z \cap F = \emptyset$; let $Z' = \{(\beta, \alpha) \mid g_{\beta,\alpha} \in Z\}$. For each $\alpha < \omega_1$, since $\{g_{\beta,\alpha} \mid \beta < \omega_1\}$ in George is homeomorphic to $\omega_1 - (\alpha + 1)$, $\{g_{\beta,\alpha} \in Z\}$ is countable. Therefore

Z' is contained in a "staircase set" S in X. If $\alpha < \beta < \omega_1$, an "$\alpha$, β-step" in X is $\{(\delta, \gamma) \in X \mid \delta < \beta$ and $\alpha \leq \gamma\}$; a "staircase set" is a union of disjoint steps. There are disjoint clopen subsets of ω_1 whose products with D separate the steps of our stair; hence there are subbasic open sets of type (b) which separate Z into disjoint countable pieces. Since George is normal, these countable discrete sets can be separated by disjoint open sets; so George is collectionwise Hausdorff.

(iv) *Not collectionwise normal.* For each $\alpha < \omega_1$, let $Y_\alpha = \{g_{\beta,\alpha} \mid \alpha < \beta < \omega_1\}$. Then $\{Y_\alpha\}_{\alpha<\omega_1}$ is a closed discrete famliy of closed sets. In order to reach a contradiction assume that $\{W_\alpha\}_{\alpha<\omega_1}$ is a family of disjoint open sets with $Y_\alpha \subset W_\alpha$.

Fix $\alpha < \omega_1$. For each $\beta \in (\alpha, \omega_1)$ there is a finite family $T_{\beta,\alpha}$ of subbasic open sets containing $g_{\beta,\alpha}$ whose intersection is contained in W_α. For each β there is a $t_{\beta,\alpha} < \beta$ such that $g_{\gamma,\alpha} \subset (\bigcap T_{\beta,\alpha})$ for all $t_{\beta,\alpha} < \gamma \leq \beta$. There is a $t_\alpha = t_{\beta,\alpha}$ for an uncountable set B_α of β. Let $S_{\beta,\alpha} = \{U \in T_{\beta,\alpha} \mid U$ is of type (c)$\}$. There is an uncountable subset C_α of B_α such that all $\beta \in C_\alpha$ have exactly the same number k_α of terms in $S_{\beta,\alpha}$.

There is a $k \in \omega$ and a countably infinite subset A of ω_1 such that $k_\alpha = k$ for all $\alpha \in A$. Choose $\beta < \omega_1$ such that $\beta > t_\alpha$ for all $\alpha \in A$. For each $\alpha \in A$ choose $\beta_\alpha > \beta$ with $\beta_\alpha \in C_\alpha$. For all $\alpha \in A$, no term of $T_{\beta_\alpha,\alpha}$ is of type (a) and all terms of $T_{\beta_\alpha,\alpha}$ of type (b) contain G_β. Just consider the restriction to G_β of $\{S_{\beta_\alpha,\alpha}\}_{\alpha\in A}$: This topology on G_α is the same as that on Bing's example G and there cannot be infinitely many disjoint basic open sets each of which is the intersection of exactly the same number of subbasic open sets. Hence we have reached a contradiction and thus proved that George is not collectionwise normal.

IX. Normality of Products

[17], [84] (1) If X is compact, infinite and metric, then $X \times Y$ is normal if and only if Y is normal and countably paracompact.

[47] (2) There is a normal space which is not countably paracompact.

[16], [85] (3) The product of a metric and a paracompact space may not be normal; but if X is metric and nondiscrete and $X \times Y$ is normal, then $X \times Y$ is countably paracompact.

[18] (4) $\beta Y \times Y$ is normal if and only if Y is paracompact.

[88], [86] (5) If X is compact and Y is normal and $w(X)$-paracompact, then $X \times Y$ is normal.

[86], [87] (6) If X is compact and $X \times Y$ is normal, then Y is $w(X)$-collectionwise normal.

[86], [87] (7) $I \times Y$ normal is not needed for the Borsuk homotopy extension theorem.

[85], [86], [87] (8) Starbird's example and the effect of closed maps.

In the 1920's it was shown that every countable product of metric spaces is metric and every arbitrary product of compact spaces is compact; but the mathematicians of this time had no natural way of determining whether the product of two normal spaces is normal. This area of topology was opened by the definition of *paracompact*. Paracompact spaces are normal but paracompactness is not preserved by products.

In 1947 Sorgenfrey [12] observed that the line with the "half open interval" topology is a paracompact Lindelöf space whose square is not normal. If intervals $[a, b)$ with $a < b$ are taken to be basic open sets for R (= real numbers), then $\{(r, -r) \mid r$ is rational$\}$ and $\{(i, -i) \mid i$ is irrational$\}$ are closed and disjoint subsets of R^2 which cannot be separated by disjoint open sets.

(1) Can there be a normal space Y such that $I \times Y$ is not normal (where I is a closed unit interval)? This question was important because $I \times Y$ normal was part of the hypothesis of Borsuk's homotopy extension theorem [90]. Also Hahn's theorem holds [17] precisely for those spaces whose product with I is normal. Dowker [17] and Katětov [84] independently gave a number of other equivalences. As a result, a normal noncountably paracompact space (or equivalently a normal space Y such that $Y \times I$ is not normal) become known as a *Dowker* space.

Suppose that X is a compact, metric, infinite space.

(a) *Suppose that Y is a normal, countably paracompact space and that H and K are disjoint closed subsets of $X \times Y$.* There is a countable basis for the topology of X

so we can let $\{U_n, V_n\}_{n\in\omega}$ be the set of all ordered pairs (U, V) of disjoint finite unions of basic open sets in X. Since X is compact, the projection map $\pi: X\times Y \to Y$ is closed. Thus

$$Y_n = Y - \pi((H - (U_n \times Y)) \cup (K - (V_n \times Y)))$$

is open in Y. Since Y is normal and countably paracompact, there is a locally finite closed refinement $\{Z_n\}_{n\in\omega}$ of $\{Y_n\}_{n\in\omega}$ with $Z_n \subset Y_n$. Then $\{(x, y) \in X \times Y \mid x \in \bigcap_{y\in Z_n} U_n\}$ and $\{(x, y) \in X\times Y \mid x \in \bigcap_{y\in Z_n} V_n\}$ are disjoint neighborhoods of H and K in $X\times Y$. So $X \times Y$ *is normal*.

(b) *Suppose that* $X\times Y$ *is normal and that* $\{U_n\}_{n\in\omega}$ *is an open cover of* Y. Choose a discrete sequence $x_0, x_1, \cdots$ converging to $x \in X$. Let $H = \{x\}\times Y$ and $K = \bigcup_{n\in\omega}(\{x_n\}\times(Y - \bigcup_{k<n}U_k))$. Since $X\times Y$ is normal and H and K are closed and disjoint, there are disjoint open subsets V and W of $X\times Y$ with $H \subset V$ and $K \subset W$. If $Z_n = \pi((\{x_n\}\times Y) - W)$ and $V_n = \pi((\{x_n\}\times Y)\cap V)$, then $V_n \subset Z_n \subset \bigcup_{k<n}U_k$, V_n is open, Z_n is closed, and $\bigcup_{n\in\omega}V_n = Y$. Therefore $\{U_n - \bigcup_{k\leq n}Z_k\}_{n\in\omega}$ is a locally finite refinement of $\{U_n\}_{n\in\omega}$. *Hence* $X \times Y$ *normal implies that* Y *is countably paracompact.*

(2) In Chapter III we discussed the construction of Dowker spaces from Souslin lines. But there is a "real" Dowker space, constructed using a Souslin tree of singular cardinality [47]. This is what one hopes set theoretic knowledge will do for you: lead you to the correct *absolute* result.

Let $F = \{f: \omega \to \omega_\omega \mid f(n) \leq \omega_{n+1}\}$.

Let $Y = \{f \in F \mid$ there is an $i \in \omega$ with $\omega <$ cofinality $(f(n)) < \omega_i$ for all $n \in \omega\}$.

Define $f < g$ $(f \leq g)$ in F provided $f(n) < g(n)$ $(f(n) \leq g(n))$ for all $n \in \omega$. Let $U_{f,g} = \{h \in Y \mid f < h \leq g\}$. Topologize Y by using $\{U_{f,g}\}$ as a basis; Y is a Dowker space. We outline a proof.

If we define $U_n = \{f \in X \mid f(i) < \omega_{i+1}$ for all $i > n\}$ then $\{U_n\}_{n\in\omega}$ is an open cover of Y with no locally finite subcover; hence Y is not countably paracompact.

To prove that Y is normal, let H and K be disjoint closed subsets of Y. For $U \subset Y$ we define the *top* of U or $t_U \in F$ by $t_U(n) = \sup\{f(n) \mid f \in U\}$. Then, by induction, we define, for each countable ordinal α, an open cover $\mathcal{G}_\alpha$ of $(H \cup K)$ by disjoint open sets having the property that:

If $\beta < \alpha < \omega_1$ and $V \in \mathcal{G}_\alpha$, then there is a $U \in \mathcal{G}_\beta$ such that:

(1) $V \subset U$.

(2) If V intersects both H and K, then $t_V \neq t_U$.

(3) If U intersects only one of H and K, then $U = V$.

Assuming such a construction, for each $\alpha < \omega_1$ and $f \in H \cup K$, let $V_{\alpha,f}$ be the (unique) term of $\mathcal{G}_\alpha$ containing f. For each $f \in H\cup K$, there must be an $\alpha_f < \omega_1$ such that $U_{\alpha_f,f}$ intersects only one of H and K since otherwise, by (1) and (2), the top of $U_{\alpha,f}$ would have to decrease infinitely many times on some coordinate. Then, by (3), $\bigcup_{f\in H}U_{\alpha_f,f}$ and $\bigcup_{f\in k}U_{\alpha_f,f}$ are disjoint open sets separating H and K and proving that Y is normal. The point is: ramification arguments are useful (not only in counting theorems).

(3) Suppose that X is a metric space. What can we say about those Y such that $X \times Y$ is normal? It is not enough [16] to require that Y be paracompact. Let X be the real line with the topology induced by adding the singleton rationals to the usual topology, and let Y be the real line with the topology induced by adding the singleton irrationals to the usual topology. Then Y is paracompact and is known as the Michael line. Although X is a complete metric space and is homeomorphic to a subset of the line, $X \times Y$ is not normal; again the rationals and irrationals on the diagonal cannot be separated.

Suppose that X is metric and nondiscrete and that Y is normal and countably paracompact. The hypotheses on Y are clearly necessary for $X \times Y$ to be normal. *We prove that, if $X \times Y$ is normal, then $X \times Y$ is countably paracompact.* From this one can prove that $X \times Y$ is normal if and only if $X \times Y$ is countably paracompact. Also if $X \times Y$ is normal, then $X \times Y$ is κ-paracompact if and only if Y is κ-paracompact and $X \times Y$ is κ-collectionwise normal if and only if Y is κ-collectionwise normal. All of the normality and paracompactness of the nonmetric factor is thus preserved in a normal product.

Assume that $X \times Y$ is normal and that $\{D_n\}_{n\in\omega}$ is a nested family of closed sets in $X \times Y$ with $\bigcap_{n\in\omega} D_n = \emptyset$. We find a family $\{K_n\}_{n\in\omega}$ of closed sets in $X \times Y$ with $D_n \cap K_n = \emptyset$ and $X \times Y = \bigcup_{n\in\omega} K_n$, thus proving that $X \times Y$ is countably paracompact.

Since X is metric, there is a family $\{\mathcal{G}_n\}_{n\in\omega}$ of locally finite open covers of X such that $\mathcal{G}_{n+1}$ refines $\mathcal{G}_n$ and $G \in \mathcal{G}_n$ implies that G has diameter less than $1/2^n$. Let $\mathcal{G} = \{G \in \bigcup_{n\in\omega}\mathcal{G}_n \mid G \text{ is not finite}\}$. For each $G \in \mathcal{G}$ choose distinct points p_G and q_G in G with $p_G \neq q_H$ for any H and G in $\mathcal{G}$.

Let π be the projection map from $X \times Y$ onto Y. For each $G \in \mathcal{G}_n$ let D_G be the closure in Y of $\pi((\overline{G} \times Y) \cap D_n)$. Then, for each $G \in \mathcal{G}$, let $P_G = \{p_G\} \times D_G$ and $Q_G = \{q_G\} \times D_G$. Define $P_n = \bigcup\{P_G \mid G \in \mathcal{G} \cap \mathcal{G}_n\}$ and $Q_n = \bigcup\{Q_G \mid G \in \mathcal{G} \cap \mathcal{G}_n\}$; let $P = \bigcup_{n\in\omega} P_n$ and $Q = \bigcup_{n\in\omega} Q_n$.

It is not difficult to prove that P and Q are closed. Since $X \times Y$ is normal there are open subsets U and V of $X \times Y$ with $P \subset U$, $Q \subset V$, and $\overline{U} \cap \overline{V} = \emptyset$. For each $G \in \mathcal{G}$ define $A_G = Y - \pi((\overline{G} \times Y) \cap V)$ and $B_G = Y - \pi((\overline{G} \times Y) \cap U)$. Note that A_G and B_G are closed in Y. Since $\mathcal{G}_n$ is locally finite, $H_n = \bigcup\{\overline{G} \times (A_G \cup B_G) \mid G \in \mathcal{G} \cap \mathcal{G}_n\}$ is closed in $X \times Y$. Also $H_n \cap D_n = \emptyset$.

Let $X^* = \{x \in X \mid x \text{ is not a limit point of } X\}$. Since $\{x\} \times Y$ is homeomorphic to Y and Y is countably paracompact, there are closed sets C_{kx} in $\{x\} \times Y$ such that $C_{kx} \cap D_k = \emptyset$ for all $k \in \omega$ and $\bigcup_{k\in\omega} C_{kx} = \{x\} \times Y$. Since $\mathcal{G}_n$ is locally finite, $J_n = \bigcup\{C_{kx} | k \leq n \text{ and } \{x\} \in \mathcal{G}_n\}$ is closed and $J_n \cap D_n = \emptyset$.

If $K_n = H_n \cup J_n$, then K_n is closed in $X \times Y$; $K_n \cap D_n = \emptyset$ and $\bigcup_{n\in\omega} K_n = X \times Y$. Thus $X \times Y$ is countably paracompact.

(4) We now turn to products with a compact factor. Dieudonné [7] observed that X compact and Y paracompact imply that $X \times Y$ is paracompact and hence normal. But $(\omega_1 + 1) \times \omega_1$ is not normal. Tamano proved [18] that $\beta Y \times Y$ *is normal (if and) only*

if Y is paracompact. The diagonal and the "end" in these spaces are the sets which are difficult to separate.

Let Y be completely regular so that βY is defined.

Assume that $\beta Y \times Y$ is normal and that $\mathcal{G}$ is an open cover of Y with no finite subcover. We prove that $\mathcal{G}$ has a locally finite open refinement.

For U open in Y, let U^* be an open set in βY with $U^* \cap Y = U$. Let $\mathcal{G}^* = \{U^* | U \in \mathcal{G}\}$ and $\mathcal{H} = \{\beta Y - U^* | U \in \mathcal{G}\}$. Since $\mathcal{H}$ has the finite intersection property, $H = \bigcap \mathcal{H} \neq \emptyset$ and $H \times Y$ is closed and disjoint from $K = \{(y, y) | y \in Y\}$. Since $\beta Y \times Y$ is normal, there is a continuous $f\colon \beta Y \times Y \to [0, 1]$ with $f(K) = 0$ and $f(H \times Y) = 1$.

For x and y in Y, define $d(x, y) = \sup\{f(t, x) - f(t, y) | t \in \beta Y\}$. For $\epsilon > 0$, let $S(\epsilon, y) = \{x \in Y | d(x, y) < \epsilon\}$ and $B(\epsilon, y) = \{b \in \beta Y | f(k, y) \leq \epsilon\}$. Since $B(\epsilon, y)$ is closed and disjoint from H, $B(\epsilon, y)$ is contained in the union of finitely many members of $\mathcal{G}^*$. Since $S(1/4, y) \subset B(1/2, y)$, $S(1/4, y)$ is contained in the union of finitely many members of $\mathcal{G}$.

If $\mathcal{J}$ is a locally finite open refinement of $\{S(1/4, y)\}_{y \in Y}$ in the metric d, then $V \in \mathcal{J}$ implies that V is open in Y and there are terms $U_1, U_2, \cdots, U_{n_v}$ of $\mathcal{G}$ whose union contains V. Thus $\{U_i \cap V | V \in \mathcal{J}$ and $1 \leq i \leq n_v\}$ is a locally finite refinement of $\mathcal{G}$.

(5) Implicit in the proof of (1)(a) is a proof that if X is compact and Y is normal and weight of $X(w(X))$-paracompact, then $X \times Y$ is normal.

For certain compact X such as $\kappa + 1$ or I^κ, the converse also holds.

Suppose that $X = Z^\kappa$ for some compact metric Z and infinite cardinal κ. Let us prove that $X \times Y$ normal implies that Y is κ-paracompact.

There is a closed subset of X homeomorphic to 2^κ. Without loss of generality we assume that κ is the smallest cardinal such that there is an open cover $\mathcal{G} = \{U_\alpha\}_{\alpha<\kappa}$ of Y of cardinality κ having no locally finite refinement. We can assume that no subset of $\mathcal{G}$ of cardinality less than κ covers Y. The assumption $2^\kappa \times Y$ normal leads to a contradiction.

Let h be the point of 2^κ such that $h(\alpha) = 1$ for all $\alpha < \kappa$. For $\alpha < \kappa$, define $X_\alpha = \{g \in 2^\kappa | g(\beta) = 0$ for all $\beta > \alpha\}$, $Y_\alpha = Y - \bigcup_{\beta<\alpha} U_\beta$, and $K_\alpha = X_\alpha \times Y_\alpha$. Then $H = \{h\} \times Y$ and $K = \bigcup_{\alpha<\kappa} K_\alpha$ are closed and disjoint in $2^\kappa \times Y$ and a proof exactly analogous to the one given for Tamano's theorem in (6) shows that $\mathcal{G}$ has a locally finite refinement.

(6) If X is the lexicographically ordered square, then $X \times \omega_1$ is normal even though ω_1 is not paracompact. However, *for all compact X, if $X \times Y$ is normal, then Y is $w(X)$-collectionwise normal.* We present Starbird's proof [86].

The weight of X is the same as the weight of $C(X)$, the set of all real valued continuous fucntions on X. By induction we can choose a family $\{f_\alpha\}_{\alpha<w(X)}$ of functions in $C(X)$ such that $\|f_\alpha\| = 1$ and, for $\alpha < \beta < w(X)$, $d(f_\alpha, f_\beta) \geq 1$.

Assume that $\{K_\alpha\}_{\alpha<w(X)}$ is a closed discrete family of closed sets in Y. Define $f\colon (X \times \bigcup_{\alpha<w(X)} K_\alpha) \to [-1, 1]$ by $f(x, y) = f_\alpha(x)$ for $y \in K_\alpha$. By Tietze's extension theorem f can be extended to $X \times Y$ if $X \times Y$ is normal.

For $y \in Y$, let $f_y \in C(X)$ be defined by $f_y(x) = f(x, y)$. Then, if $U_\alpha = \{y \in Y |$

$d(f_y, f_\alpha) < ½\}$, $\{U_\alpha\}_{\alpha<w(X)}$ is a family of disjoint open sets in Y with $K_\alpha \subset U_\alpha$. Thus $X \times Y$ is collectionwise normal.

(7) Starbird [86] uses the technique of (6) together with the technique of the usual proof of Tietze's extension theorem to prove the following powerful theorem (also proved independently by Morita [89]).

If X is compact, then $C(X)$ is an AE ($w(X)$-collectionwise normal). AE stands for absolute extensor which means that if K is a closed subset of a $w(X)$-collectionwise normal space Y and $f\colon K \to C(X)$ is continuous then f can be extended to Y.

This theorem can be used to prove the following generalization of Borsuk's homotopy extension theorem *without* the traditional hypothesis that $I \times Y$ be normal. The absence of this theorem was one of the main reasons that normality of products became important in the first place!

If K is a closed subset of a normal space Y and $f\colon (I \times K) \cup (\{0\} \times Y) \to [0, 1]$ is continuous, then there is an extension of f to $I \times Y$.

To prove this theorem from the previous one observe that, since Y is normal, Y is ω_0-collectionwise normal and hence $f \restriction (I \times K)$ can be extended to all of $I \times Y$ by a function F. Although f may not agree with F on $\{0\} \times Y$, they differ by a simple continuous function and the function F can be linearly modified to f on $(0, y)$, $F(0, y)$ on $(½\,|\,F(0, y) - f(0, y)\,|\,, y)$ and $F(1, y)$ on $(1, y)$.

(8) Closed maps preserve normality, paracompactness, κ-paracompactness and κ-collectionwise normality. The theorems presented in (3), (5), and (6) are used to prove that if X is a metric or compact space (or the product of a metric space and a compact space), then the class of all spaces whose product with X is normal (or paracompact or κ-paracompact or κ-collectionwise normal) is closed under closed maps [85], [86], [87]. However, if G is Bing's space described in Chapter VIII(7), *there is a metric space Y and a closed image Z of Y such that $G \times Y$ is normal, but $G \times Z$ is not normal.*

Define $Y = D \times (\omega + 1)$ where D is ω_1 with the discrete topology. Define Z to be the quotient space of Y obtained by identifying all points of $D \times \{\omega\}$ to a single point p.

Since G is normal and countably paracompact and $\omega + 1$ is compact and metric, $G \times (\omega + 1)$ and $G \times Y$ are normal.

Define $H = \{f_\alpha\}_{\alpha<\omega_1} \times \{p\}$. For each $\alpha < \omega_1$ reorder $\{f_\beta\}_{\beta<\alpha}$ as $\{g_{\alpha i}\}_{i\in\omega}$; then define $K = \{(g_{\alpha i}, (\alpha, i)) \in G \times Z \mid \alpha < \omega_1 \text{ and } i < \omega\}$. The sets H and K are closed and disjoint in $G \times Z$.

Let U be an open set in $G \times Z$ which contains H. We find a point in $K \cap \overline{U}$, thus proving that $\mathcal{G} \times Z$ is not normal.

Since $(f_\alpha, p) \in U$ for all $\alpha < \omega_1$, there is an uncountable subset S of ω_1, a $k \in \omega$, a sequence $e_{0\alpha}, e_{1\alpha}, \cdots, e_{k\alpha}$ of 0's and 1's a sequence $A_{0\alpha}, A_{1\alpha}, \cdots, A_{k\alpha}$ of subsets of ω_1, and a neighborhood N_α of p in Z such that $(\bigcap_{n\le k} U_{A_{n\alpha}, e_{n\alpha}} \times N_\alpha) \subset U$. We can choose an $i \le k + 1$ and an uncountable subset T of S such that, for all

$\alpha < \beta < \omega_1$, $A_{n\alpha} = A_{n\beta}$ for $n < i$ and $A_{n\alpha} \neq A_{j\beta}$ for $i \leq j \leq n$.

Let M be an infinite countable subset of T and choose $\beta \in T$ with $\beta > \alpha$ for all $\alpha \in M$. There is a $j \in \omega$ and an uncountable subset R of T such that $N_\alpha \supset \{(\beta, m) \mid m > j\}$ for all $\alpha \in R$. Choose $m > j$ such that, for some $\gamma \in M$, $f_\gamma = g_{\beta,m}$; then $(g_{\beta,m}, (\beta, m)) \in K \cap \overline{U}$.

X. Box Products

Suppose that, for each $n \in \omega$, X_n is a nondiscrete space, and X is the box product of $\{X_n\}_{n \in \omega}$. Then:

[91] (1) X is not connected.

[91] (2) X is not first countable.

[91] (3) X is not compact.

[98], [99] (4) X need not be normal even if all X_n are separable and metric.

[95] (5) X need not be normal even if all X_n are compact.

[92], [95] (6) CH implies that X is paracompact if all X_n are compact and either of weight $\leq c$ or scattered.

[93], [96] (7) GCH decides if X is normal (paracompact) if all X_n are ordinals.

[94], [95], [96] (8) MA decides if X is normal if all X_n are compact and first countable.

If $\{X_\alpha\}_{\alpha \in A}$ is a family of nondiscrete spaces, then the box product X of $\{X_\alpha\}_{\alpha \in A}$ is $\Pi_{\alpha \in A} X_\alpha$ topologized by taking the set of all boxes in $\{X_\alpha\}_{\alpha \in A}$ as a basis. A *box* in $\{X_\alpha\}_{\alpha \in A}$ is a product $\Pi_{\alpha \in A} U_\alpha$ where each U_α is open in X_α. To the uninitiated this seems like the natural definition for the product topology and it *is* the product topology when A is finite.

The box product of Hausdorff spaces, regular spaces, and completely regular spaces are Hausdorff, regular and completely regular, respectively. But few other elementary properties are preserved by box products. If X is a box product of ω closed unit intervals, for instance, X has uncountable weight, density, spread, cellularity, Lindelöf number, and character, and X is not connected. Is this X normal or paracompact? We now know that Martin's axiom implies that X is paracompact, but without set theoretic assumptions we do not know the anwser; we have *no* positive nonconsistency results. And we have no positive consistency results for box products of uncountably many nondiscrete spaces.

So why bother if this is such a bad topology? First of all, the concept is a natural set theoretic one and it presents difficulties in precisely the area (between normality and paracompactness) where so many basic abstract space problems are hard. Second, it has recently been used to construct counterexamples for a number of well-known conjectures ([47], [99]). It is useful as an example machine particularly for normality problems.

Box products are considered in Bourbaki and Kelley and in a variety of places;

[91] is a good place to find the basic theorems together with Stone's conjecture that the box product of uncountably many lines is normal. (This conjecture is still untouched.) Tamano became interested in box products as a tool for attacking Dowker's problem and I used this idea to solve Dowker's problem (Chapter IX(2)). Our enthusiasm for box products led us to make overoptimistic conjectures. Tamano actually announced a proof that any box product of metric spaces is normal (disproved by Van Douwen [98]). I conjectured that any box product of compact spaces is paracompact (disproved by Kunen [95]). But basically this is a success story: four years ago no one knew anything about normality in box products. I broke the ice with a couple of messy papers [92], [93] which are now only of historical interest; better theorems and better proofs are found in Kunen's pretty papers [95], [96]. Van Douwen and Vaughan also have nice papers [98], [99]. We begin to understand.

Throughout this chapter we assume that X is the box product of $\{X_n\}_{n\in\omega}$ and that each X_n is nondiscrete.

(1) If $x \in X$, let $E(x) = \{y \in X \mid y(n) \neq x(n)$ for at most finitely many $n\}$. We prove that $E(x)$ *contains the component of* X *containing* x. Since each X_n has at least two points $E(x) \neq X$ and X is thus not connected.

Suppose that $p \in X - E(x)$. For each n and k in ω, choose an open subset $U_{kn}(x)$ of X_n such that $x(n) \in U_{kn}(x)$, $p(n) \notin U_{kn}(x)$ unless $p(n) = x(n)$, and $U_{kn}(X) \supset \overline{U_{(k+1)n}(x)}$. Let $Y = \{y \in X \mid$ for some $k \in \omega$ and for infinitely many $n \in \omega$, $y(n) \notin U_{kn}\}$; $p \in Y$ but $x \notin Y$. Since Y is clopen p does not belong to the component of X containing x.

(2) Suppose that $x \in X$ and that $x(n)$ is a limit point of X_n for each n. Suppose that, for each $k \in \omega$, $V_k = \Pi_{n\in\omega} V_{kn}$ is a box in X containing x such that $x \in V_k$. Assume that $V_{kn} \subsetneqq \bigcap_{i<k} V_{in}$. Then $V = \Pi_{n\in\omega} V_{nn}$ is open in X, contains x, and is not contained in V_k for any $k \in \omega$. Thus X *is not first countable.*

(3) Suppose that, for each $n \in \omega$, $x_{n0}, x_{n1}, \cdots$ is a discrete sequence of distinct points in X_n. For $k \in \omega$, define $x_k \in X$ by $x_k(n) = x_{nk}$. Since $\{x_k\}_{k\in\omega}$ has no limit point in X, X *is not compact.*

(4) *Let* X_0 *be the set of all irrational numbers with the usual topology. Let* $X_n = \omega + 1$ *for all* $n > 0$. We now prove Van Douwen's theorem that this X *is not normal.*

Let Y be the box product of $\{X_n\}_{n>0}$ and Y^* the ordinary product. Let $W = \{y \in Y \mid y(n) < \omega$ for all $n \in \omega\}$ as a subspace of Y and let W^* be this set as a subspace of Y^*. We think of X as $W^* \times Y$ since W^* is homeomorphic to the irrationals.

Let $H = \{(w, w) \in W^* \times Y \mid w \in W\}$.

Let $K = W^* \times (Y - W)$.

Then H and K are trivially closed and disjoint. Suppose that $K \subset V$ which is open in X; we show that $H \cap \overline{V} \neq \emptyset$ which proves that X is not normal.

For each $n \in \omega$, choose $f_n \in \omega^\omega$ by induction on n as follows. If f_i has been chosen for all $i < n$, define

$$w_n = \left(f_0(0), \cdots, \sum_{i<n} f_i(n-1), 0, 0, \cdots\right) \quad \text{and}$$

$$y_n = \left(f_0(0), \cdots, \sum_{i<n} f_i(n-1), \omega, \omega, \cdots\right).$$

Since $(w_n, y_n) \in K$, there is $f_n \in \omega^\omega$ such that

$$\{w_n\} \times \left(\{f_0(0)\} \times \cdots \times \left\{\sum_{i<n} f_i(n-1)\right\} \times [f_n(n), \omega] \times [f_{n+1}(n), \omega] \times \cdots\right) \subset V.$$

If $x = \Pi_{n\in\omega}\{\Sigma_{i\leq n} f_i(n)\}$, then $(x, x) \in H \cap \overline{V}$.

Van Douwen observes that $Y - W$ *is not a* G_δ *set in* Y (the same argument given in (2)) and a corollary of this fact is that Y *is not hereditarily normal.* This follows from the fact that $(\omega + 1) \times Y$ is homeomorphic to Y and an old theorem of Katětov [84] which states that if Y is not perfectly normal and if Z has a nondiscrete countable subset, then $Z \times Y$ is not hereditarily normal.

A related example is given by Vaughan to show that *there is a metrizable space* Z *and an* ω_1*-metrizable space* X *such that* $Z \times X$ *is not normal.* For each $n \in \omega$, let $Z_n = \omega_1$ with the discrete topology and let $X_n = \omega_1 + 1$ with the topology obtained by adding the set of all subsets of ω_1 to the usual topology. Define Z to be the usual product of $\{Z_n\}_{n\in\omega}$ and define X to be the box product of $\{X_n\}_{n\in\omega}$. Clearly Z is metrizable and X is ω_1-metrizable.

To see that $Z \times X$ is not normal define $H = \{(z, z) \in Z \times X \mid z \in Z\}$ and $K = Z \times (X - Z)$. The same argument given above shows that every open set containing K has a limit point in H and $Z \times X$ is hence not normal.

(5) *For each* $n \in \omega$, *let* $X_n = 2^{(c^+)}$; *again* X *is not normal.*

We shall prove in (6) that for a countable box product of compact spaces to fail to be normal without set theoretic assumptions, the factors must have weight greater than c. Thus, this is the simplest possible example.

Kunen proves that the diagonal $D = \{x, x, \cdots \in X\}$ is not normal; hence X is not normal. He uses three ideas.

(a) D is homeomorphic to $2^{(c^+)}$ with the G_δ topology. The mapping d defined by $d(x, x, \cdots) = x$ is the desired homeomorphism.

(b) We think of D as 2^A with the G_δ topology where $A = (c^+ \times \omega) \cup [c^+]^2$; any set A of cardinality c^+ would do.

(c) Let $A_\gamma = (\gamma \times \omega) \cup [\gamma]^2$ and let $\pi_\gamma\colon 2^A \to 2^{A_\gamma}$ be the projection map. Engelking and Kartowicz [100] prove that, if U and V are disjoint open subsets of 2^A with the G_δ topology, then there is a $\gamma \in C^+$ such that $\pi_\gamma(U) \cap \pi_\gamma(V) \neq \emptyset$.

If $f \in 2^A$ and $\alpha < c^+$, let $f_\alpha\colon \omega \to 2$ be defined by $f_\alpha(n) = f(\alpha, n)$.

Define $H = \{f \in 2^A \mid \alpha < \beta < c^+$ and $f_\alpha = f_\beta$ imply that $f(\{\alpha, \beta\}) = 0\}$.

Define $K = \{f \in 2^A \mid \alpha < \beta < c^+$ implies that $f(\{\alpha, \beta\}) = 1\}$.

Then H and K are closed and disjoint since there cannot be c^+ distinct f_α's. But if $\gamma < c^+$, there *are* γ distinct f_α's so $\pi_\gamma(H) \cap \pi_\gamma(K) \neq \emptyset$ for all γ. Thus, by (c), H and K cannot be separated and D and X are thus not normal.

(6) We now give Kunen's proof [95] that *CH implies that* X *is paracompact if each* X_n *is compact and of character* $\leq c$. When you study the proof, it is not difficult

to see that compact can be replaced by locally compact and paracompact. So Kunen's theorem generalizes the theorem proved in [92] where each X_n is assumed to be locally compact and metric (and σ-compact).

Kunen uses four lemmas. The first two hold for arbitrary spaces X and Y having nothing to do with box products, and they can be proved using Michael's theorems and techniques [13], [14], [15]. If κ is a cardinal, a space Y is *κ-open* if every intersection of less than κ-open sets is open; Y is κ-Lindelöf if every open cover of Y has a subcover of cardinality less than or equal to κ.

(A) If Y is κ-open and κ-Lindelöf, then Y is paracompact.

(B) If $f: X \to Y$ is a closed map, Y is paracompact, and $f^{-1}(y)$ is Lindelöf for all $y \in Y$, then X is paracompact.

Suppose that each X_n is a compact space. Again, for $x \in X$, define $E(x) = \{y \in X \mid y(n) \neq x(n)$ for at most finitely many $n\}$. Let $Y = \{E(x) \mid x \in X\}$ be topologized by declaring U to be open in Y if $\bigcup U$ is open in X. Kunen observes the following.

(C) Y is ω_1-open.

(D) E is a closed map from X onto Y.

If CH and each X_n has weight $\leq c$, then Y has weight $\leq \omega_1$; hence Y is ω_1-Lindelöf. So, by (C) and (A), Y is paracompact. Since X_n is compact $E(x)$ is σ-compact for each $x \in X$; hence each $E(x)$ is Lindelöf. So, by (D) and (B), X is paracompact.

Let us now prove Kunen's theorem that *CH implies that, if each X_n is compact and scattered, then X is paracompact*. Exactly the same proof works when we prove that *this X is c-Lindelöf*. This implies that Y is c-Lindelöf.

If Z is a space, by Z' we mean the subspace of Z consisting of the nonisolated points. For each ordinal α, we define Z^α by induction. $Z^0 = Z$ and, for limit α, $Z^\alpha = \bigcap_{\beta<\alpha} Z^\beta$. If $\alpha = \beta + 1$ we define $Z^\alpha = (Z^\beta)'$. If Z is a compact scattered space, then there is a smallest α with $Z^\alpha = \emptyset$; $\alpha = \beta + 1$ and we call β the rank of Z; Z^β is finite.

Suppose that each X_n is scattered and compact and that $\mathcal{G}$ is an open cover of X. We find a closed refinement of $\mathcal{G}$ covering X.

For each $\alpha < \omega_1$, $n \in \omega$, and $f \in c^\alpha$, we choose a nonempty closed subset $K_n(f)$ of X_n by induction as follows. Define $K_n(\emptyset) = X_n$ (if $f \in c^0$, $f = \emptyset$). If β is a limit ordinal and $f \in c^\beta$ and $K_n(f \upharpoonright \alpha)$ has been defined for all $n \in \omega$ and $\alpha < \beta$, then define $K_n(f) = \bigcap_{\alpha<\beta} K_n(f \upharpoonright \alpha)$. Suppose that $f \in c^\alpha$ and that $K_n(f)$ has been defined for all $n \in \omega$. Let γ_n be the rank of $K_n(f)$. Let H and K be the subspaces of X consisting of $\Pi_{n\in\omega} K_n(f)^{\gamma_n}$ and $\Pi_{n\in\omega} K_n(f)$, respectively. By CH, $|H| \leq c$. So we can find a closed cover $\{H_\delta\}_{\delta<c}$ of K such that, for each $\delta < c$ and $n \in \omega$, there is a closed subset $H_{\delta n}$ of $K_n(f)$ with

(a) $H_\alpha = \Pi_{n\in\omega} H_{\delta n}$,

(b) if $y \in H \cap H_\delta$, there is a $U \in \mathcal{G}$ with $H_\delta \subset U$, and

(c) if there is no $U \in \mathcal{G}$ with $H_\delta \subset U$, then there is $n \in \omega$ with the rank of $H_{\delta n}$ less than the rank of $K_n(f)$. If $\gamma = \alpha + 1$, $g \in c^\gamma$, and $g \upharpoonright \alpha = f$, define $K_n(g) = H_{g(\alpha)n}$.

If $x = x_0, x_1, \cdots \in X$, then, for each $\alpha < \omega_1$, we can choose $f_{x\alpha} \in c^{\alpha}$ by induction such that $f_{x\alpha} \upharpoonright \beta = f_{x\beta}$ for all $\beta < \alpha$ and $x \in K_{\alpha} = \Pi_{n\in\omega} K_n(f_{x\alpha})$. Since there is no infinite decreasing sequence of ordinals, by (c), there is an $\alpha(x) \in \omega_1$ such that $K_{\alpha(x)}$ is contained in a member of $\mathcal{G}$. Since $|\bigcup_{\alpha<\omega_1} c^{\alpha}| = c$, $|\{K_{\alpha(x)}\}_{x\in X}| \leq c$. So $\mathcal{G}$ has a (closed) subcover of cardinality at most c and we have proved that X is c-Lindelöf.

(7) *Suppose that each* X_n *is an ordinal* α_n. *Let* $S = \{n \in \omega \mid \mathrm{cf}(\alpha) > \omega\}$.

(a) *If* $S = \emptyset$, *CH implies that* X *is paracompact if each* α_n *is* σ*-compact.*

(b) *If* $S \neq \emptyset$, X *is not normal unless there is a regular cardinal* κ *such that either* (i) S *has exactly one element cofinal with* κ, *and* $\alpha_n < \kappa$ *for all* $n \in (\omega - S)$, *or* (ii) $\alpha_n = \kappa$ *for all* $n \in S$, *and* $\alpha_n < \kappa$ *for all* $n \in (\omega - S)$.

(c) *If* $S \neq \emptyset$, (i) *or* (ii) *holds, and* κ *is not the successor of a cardinal cofinal with* ω, *then GCH implies that* X *is not normal.*

(d) *If* $S \neq \emptyset$, (i) *or* (ii) *holds, and* $\kappa = \lambda^+$ *for some* λ *with* $\mathrm{cf}(\lambda) = \omega$, *then GCH implies that* X *is normal if and only if, for some* $\beta < \lambda$, *all but finitely many* α_n *are less than* β.

This theorem completely decides modulo GCH whether a given product of ordinals is normal or not. One should observe that without loss of generality all σ-compact ordinals can be assumed to be compact since a σ-compact ordinal is the discrete union of compact ordinals.

(a) follows immediately from the fact that ordinals are scattered as we proved in (6).

(b) follows from the fact that, if β and γ are ordinals and $\mathrm{cf}(\beta) > \omega$, then $\mathrm{cf}(\beta) \times (\mathrm{cf}(\beta) + 1)$ is not normal (hence $\beta \times \gamma$ is not normal).

(c) and (d) are proved by looking at the box product Y of $\{\beta_n\}_{n\in\omega}$ where $\beta_n = \alpha_n$ if α_n is σ-compact and $\beta_n = \alpha_n + 1$ if $\mathrm{cf}(\alpha_n) > \omega$ and $\beta_n = \alpha_n$ if $\mathrm{cf}(\alpha_n) \leq \omega$ (recall that we can assume that α_n is compact if $\mathrm{cf}(\alpha_n) = \omega$).

(c*) H *and* K *are closed and disjoint in* X *implies that* H *and* K *are closed in* Y.

By (a), CH implies that Y is normal. So, if (c*), then CH implies that X is normal.

(c**) *Either* (x) S *is finite and* $|\Pi\{\beta_n \mid n \notin S\}| < \kappa$, *or* (xx) S *is infinite and* $\lambda^{\omega} < \kappa$ *for all* $\lambda < \kappa$.

Certainly GCH and the hypothesis of (c) yield (c**).

Suppose that $F \subset \omega$ is finite, $\kappa = \lambda^+$, $\mathrm{cf}(\lambda) = \omega$, $\beta < \lambda$, and $\beta_n < \beta$ for all $n \in F$. Since $\Pi_{n\in F-S}\beta_n$ is compact, we can assume that $F \subset S$. Thus GCH implies that (c**) holds in the positive half of (d).

Let us now prove that (c**) *and either* (b)(i) *or* (b)(ii) *implies* (c*). This proves (c) and the positive half of (d).

Suppose that H and K are disjoint closed subsets of X and that f belongs to the closures in Y of both H and K. Assume that (b)(i) or (b)(ii). Let $M = \{n \in S \mid f(n) = \beta_n\} \neq \emptyset$ and let β be the β_n for all $n \in M$.

By (c**), $|\Pi_{n\in(\omega-M)}\beta_n| = \lambda < \kappa$. So we can index $\Pi_{n\in\omega-M} f(n) = \{g_{\gamma}\}_{\gamma<\lambda}$ where each $g \in \Pi_{n \in (\omega - M)} f(n)$ is g_γ for λ different γ. By induction, for each $\gamma < \lambda$, choose $h_\gamma \in H$ and $k_\gamma \in K$ which are contained in the open set $\Pi_{n \in M}(\alpha_n - \bigcup_{\delta<\gamma}\lambda_\delta) \times \Pi_{n \in (\omega - M)}((f(n)+1) - g_\gamma(n))$ where $\lambda_\beta = \bigcup_{n \in M} h_\beta(n) \cup k_\beta(n)$. Let $\lambda = \sup\{\lambda_\gamma\}_{\gamma<\lambda}$. Since $\lambda < \kappa$, the point $g \in Y$ defined by $g(n) = \lambda$ for $n \in M$ and $g(n) = f(n)$ for $n \in \omega - M$ belongs to X and to the closures of both H and K. Thus (c*) holds.

We now turn to a proof of the negative half of (d). Suppose that $\alpha_0 = \lambda^+$ for some λ with $\mathrm{cf}(\lambda) = \omega$ and that $\sup\{\beta_n\}_{n>0} = \lambda$. Each of the spaces we are considering has a closed subset homeomorphic to this X. We can assume that each $\beta_n = \gamma_n + 1$ where γ_n is a regular cardinal. GCH implies that there is a λ^+-scale in $\Pi_{n\in\omega}\gamma_n$. Kunen proves:

(d*) *If there is a λ^+-scale in $\Pi_{n\in\omega}\gamma_n$, then X is not normal.*

If $\{f_\alpha\}_{\alpha<\lambda^+}$ is the scale let

$$H = \{h \in X \mid h(n) = \beta_n \text{ for all } n > 0\} \quad \text{and}$$
$$K = \{k \in X \mid k(n) = f_{k(0)}(n) \text{ for all } n > 0\}.$$

These disjoint closed subsets of X cannot be separated.

(8) We now turn to Martin's axiom and ask about box products under this assumption.

(a) *Suppose that each X_n is a compact, first countable space. Then Martin's axiom implies that X is paracompact.*

We have already shown that every dominating family of functions from ω into ω has cardinality at least c. An immediate consequence of this is the fact that X is c-open. By Arhangelskiĭ's theorem (II(7)), each X_n has weight $\leq c$. So the above theorem follows from the argument given in (6).

(b) Suppose that $0 \leq i \leq j < \omega$, that X_n is an ordinal α_n of uncountable cofinality for $0 \leq n < i$, that X_n is a σ-compact ordinal α_n for $i \leq n < j$, and that X_n is a countable ordinal α_n for $j \leq n$.

We can assume that σ-compact ordinals are compact. Thus we can assume that $i = j$ since the normality of X is not affected by the compact factor $\Pi_{i\leq n<j} X_n$.

If $i = 0$, Martin's axiom implies that X is paracompact by (a).

If $i \neq 0$, by (7)(b), X is not normal unless there is a cardinal κ such that either $i = 1$ and $\mathrm{cf}(\alpha_0) = \kappa$ or $\alpha_n = \kappa$ for all $n < i$.

If $\kappa = c$, assuming Martin's axiom, there is a κ-scale, so by (7)(d*), *X is not normal.*

If $c < \kappa$, then X is normal. For if $c < \kappa$, $|\Pi_{n>j}\alpha_n| = c < \kappa$, so (7)(c**) holds.

Let us prove that, *if* $\kappa < c$ and Martin's axiom holds, *then X is again normal.*

Let P be the box product of $\{\alpha_n\}_{n>j}$. *If $p \in P$ and $\{U_\beta\}_{\beta<\kappa}$ is any family of neighborhoods of p, then Martin's axiom implies that there is a family $\{V_n\}_{n\in\omega}$ of neighborhoods of p such that, for all $\beta < \kappa$, there is an n with $V_n \subset U_\beta$.*

Let $Q = \Pi_{n<i}\alpha_n$. Q is normal, κ-Lindelöf, and countably compact. So, since P has the property given above, *the projection map $\pi: Q \times P \to P$ is closed.* To see this observe that, if K is closed in $Q \times P$ and $p \in P$, for each $q \in Q$ there are

neighborhoods N_q of q and U_q of p with $(N_q \times U_q) \cap K = \emptyset$. There is a subset Q' of Q such that $|Q| = \kappa$ and $\{N_q\}_{q \in Q'}$ covers Q. Thus there is a set $\{V_n\}_{n \in \omega}$ of neighborhoods of p such that, for $q \in Q'$, there is an n with $V_n \subset U_q$. Let $M_n = \bigcup\{N_q \mid V_n \subset U_q\}$. Since Q is countably compact there is a $k \in \omega$ with $\bigcup_{n<k} M_n = Q$ and $p \in \bigcap_{n<k} V_n$. Hence $\bigcup_{n<k} M_n \times V_n$ contains $Q \times \{p\}$ but does not intersect K.

Since π is closed, P is paracompact and Q is normal and $Q \times P$ is thus normal.

Problems

The following problems are unsolved so far as I know. They are being solved almost daily, of course, for they are problems which people are working on. Some are very hard, basic, long unsolved and frequently worked on problems; others are just things someone ran across and did not know the answer to. The names by the problem are not those of the first person to ask the problem or even the person currently most actively working on the problem: the name implies that that person once mentioned this problem to me and probably can fill in anyone interested in the problem on more details and background.

All spaces are assumed to be Hausdorff.

A *map* is a function which is continuous and onto.

A. Cardinal Function Problems

Mrowka 1. If every zero set is in B (clopen sets), then is every zero set the intersection of a countable number of clopen sets? B (clopen sets) is the smallest family containing the clopen sets which is closed under countable unions, countable intersections and complements.

Juhász and Hajnal 2. Is there a regular space X with cardinality greater than c which is not hereditarily separable but every closed subset is separable?

J. and H. 3. If X is an infinite space and the number of open sets in X is m, then is $m^{\omega} = m$?

J. and H. 4. Does every Lindelöf space of cardinality ω_2 contain a Lindelöf subspace of cardinality ω_1?

J. and H. 5. If X is hereditarily separable and compact (subset of 2^{ω_1}), then is $|X| \leq c$?

J. and H. 6. If X is a regular space of countable spread, does $X = Y \cup Z$ where Y is hereditarily separable and Z is hereditarily Lindelöf?

Arhangelskiĭ 7. If X is a regular Lindelöf space each point of which is a G_δ, then is $|X| \leq c$?

A. 8. If a hereditarily normal space X has countable cellularity and tightness, is $|X| \leq c$? (No if $V = L$ without hereditary normality.)

The *tightness* of X at x is $t(X, x) = \min\{m \mid \forall A \subset X \text{ with } x \in \overline{A},\ \exists B \subset A \text{ with } |B| \leq m \text{ and } x \in \overline{B}\}$. The tightness of X is $\sup\{t(X, x) \mid x \in X\}$.

A. 9. Does each compact hereditarily normal space of countable tightness contain a converging (nontrivial) sequence of points? have a point of countable character? (No if $V = L$ without hereditary normality.)

A. 10. Does every compact homogeneous space with countable tightness have cardinality $\leq c$?

Smirnov 11. Does every hereditarily normal compact space contain a point with a countable Δ-base?

A Δ-*base* for a point x is a family B of open sets such that every neighborhood of x contains a member of B having x in its closure.

Ponomarev 12. Is a compact space of countable tightness a sequential space? (No if $V = L$.)

X is *sequential* provided $Y \subset X$ implies that Y is closed whenever for all y_1, $y_2, \cdots$ in Y with $y_n \to y$, then $y \in Y$.

13. Is the product of two Lindelöf spaces c-Lindelöf?

14. Is every separable metric space, such that every nowhere dense closed subset is σ-compact, σ-compact?

Purish 15. Is orderable equivalent to monotonically normal for compact, separable, totally disconnected spaces?

X is *monotonically normal* if there is function f which assigns to each pair (H, K) of disjoint closed subsets of X an open set $f(H, K)$ such that $H \subset f(H, K)$, $\overline{f(H, K)} \subset X - K$, and $H \subset H'$ and $K \subset K'$ imply that $f(H, K) \subset f(H', K')$.

Telgarski 16. Is there a compact space X with no isolated points which does not contain a zero-dimensional closed subset with no isolated points? (No if $V = L$.)

T. 17. Is every image of a scattered space under a closed map scattered?

Semadeni 18. Is every scattered completely regular space zero-dimensional?

Hajnal 19. Suppose A and B are sets with $|A| = 2^{\omega_1}$ and $|B| = 2^{\omega_0}$. Color $A \times B$ with two colors. Must there be $A' \subset A$ and $B' \subset B$ such that $|A| = \omega_0$, $|B| = \omega_1$, and $A' \times B'$ is one color? (*Yes* is "consistent".)

Stephenson 20. Is the property *initially m-compact* productive for regular uncountable m? (Every open cover of cardinality $\leq m$ has a finite subcover if X is initially m-compact.)

Morita 21. Is every normal space X countably compactifiable? (Is X dense in a countably compact space S such that every countably compact subset of X is closed in S?)

B. Souslin and Compactness Problems

Kunen 1. If X is ccc and Y is ccc, but $X \times Y$ is not ccc, then is there a Souslin line?

2. Is there a Souslin line if there is a normal noncountably paracompact space (a Dowker space) which is one (or many) of the following? (Yes if CH.)

first countable

separable

cardinality ω_1

ccc

real compact

monotonically normal (See 15.)

Smirnov 3. Does every compact space contain either a copy of $\beta N - N$ or a point of countable Π-character?

A Π-base for a point x is a family B of open sets such that every neighborhood of x contains a member of B.

S. 4. Is there a ccc, compact space X with countable Π-character (or with $|X| \leq c$) which is not separable?

Tall 5. Is every compact space supercompact? (X is supercompact if there is a subbasis S for the closed sets such that if $T \subset S$ and every two members of T meet, then $\bigcap T \neq \emptyset$.)

T. 6. A space is Baire if no open set is first category. Is there a Baire space whose square is not Baire? (Yes with Martin's axiom even for metric spaces.)

Henriksen 7. Is the set of remote points in βR dense in $\beta R - R$? (R is the real line and p is remote in βR if p is not in the closure of any nowhere dense subset.)

Frolík 8. Is there a P-point in $\beta N - N$? (Yes if Martin's axiom.)

Arhangelskiĭ 9. Does every hereditarily separable compact space have a point of countable character? a nontrivial converging sequence? a butterfly point? (A point x for which there are closed subsets A and B with $\{x\} = \overline{A - \{x\}} \cap \overline{B - \{x\}}$ is a butterfly point.)

Saks 10. Is there a product of compact topological groups which is not countably compact?

Hager 11. If X is a dense subset of a compact Y and every open set containing X is C^*-embedded in Y, then is X C^*-embedded in Y?

Arhangelskiĭ 12. Is there an infinite homogeneous extremally disconnected compact space? (Yes if CH.)

Blaire 13. If X is Lindelöf and Y is real compact, does X closed in $X \cup Y$ imply that $X \cup Y$ is real compact?

Kunen 14. Can a compact space be decomposed into more than c closed G_δ's?

Bandy 15. Are there two normal countably compact spaces whose product is not countably compact?

C. Separable-Lindelöf Problems

Juhász and Hajnal 1. Is there a first countable, ccc, density $\leq c$ space with uncountable spread? (Yes if CH or a Souslin line.)

J. and H. 2. Is there a regular space with cardinality greater than c which has countable spread? (Consistent with CH that the answer be yes.)

J. and H. 3. Is there a regular, hereditarily Lindelöf space with weight greater than c? (Consistent with CH that the answer be yes.)

J. and H. 4. Is there a (regular) hereditarily separable space X with $|X| > 2^{\omega_1}$?

J. and H. 5. Is there a regular space which is hereditarily separable but not Lindelöf, or vice versa? (Yes in both cases if CH or a Souslin line.)

J. and H. 7. Is the density $\leq$ the smallest cardinal greater than the spread for compact space? regular spaces? regular hereditarily Lindelöf spaces?

J. and H. 8. Could a compact hereditarily separable space have cardinality greater than c?

D. Metrizability Problems

Jones 1. Is there a normal nonmetrizable Moore space? (Yes if (MA + $\neg$CH).)

Alexandroff 2. Is there a normal nonmetrizable image of a metric space under a compact open map? (the metacompact normal Moore space problem) (Yes if (MA + $\neg$CH).)

Fleissner 3. Is there a first countable, normal, collectionwise Hausdorff space which is not collectionwise normal? (Yes if (MA + $\neg$CH).)

Reed 4. Is every countably paracompact Moore space normal?

R. 5. Is there a countably paracompact Moore space which is not paracompact? (Yes if (MA + $\neg$CH).)

Wilder 6. Is every perfectly normal manifold metrizable? (Yes if $\diamondsuit$.)

Zenor 7. Is every perfectly normal manifold subparacompact?

X is *subparacompact* if every open cover of X has a σ-locally finite closed refinement. Problem 6 = problem 7. (Yes if $\diamondsuit$.)

Hodel 8. Is every perfectly normal collectionwise normal space paracompact? (No if $\diamondsuit$ or (MA + $\neg$CH).)

H. 9. Is every perfect (closed sets are G_δ) space θ-refinable? (No if $\diamondsuit$.)

X is *weak θ-refinable*: Every open cover of X has a refinement $R = \bigcup_{n\in\omega}\{R_n\}$ such that, for all $x \in X$, there is an $n \in \omega$ such that $\{U \in R_n \mid x \in U\}$ is finite but not $\emptyset$. X is *θ-refinable* if each R_n covers X.

X is *quasi-developable* if there is a family $\{G_n\}_{n\in\omega}$ of collections of open sets such that, for all $x \in X$ and neighborhood U of x, there is an $n \in \omega$ such that $U \supset \bigcup\{V \in G_n \mid x \in V\} \supset \{x\}$. X is developable if each G_n covers X.

X is quasi-metrizable if $d(x, y) \neq d(y, x)$ is possible in the "metric".

H. 10. Is every normal space with a point countable base metrizable? (No if (MA + $\neg$CH).)

H. 11. Is every perfectly normal space with a point countable base metrizable? (Add paracompact or collectionwise normal to the hypothesis if you wish.) *Ponomarev*: Replace *perfectly normal* in this question with *regular* Lindelöf or *regular hereditarily* Lindelöf. (All answers here are yes if there is a Souslin line.)

Nyikos 12. Is there a perfectly normal non-Archimedean space which is not metrizable? (Yes if (MA + $\neg$CH) or there is a Souslin line.)

X is non-Archimedean if there is a basis B for the topology of X such that U and V in B imply that either $U \supset V$, $V \supset U$ or $U \cap V = \emptyset$.

Zenor 13. Is every countably compact space with a G_δ-diagonal metrizable?

Arhangelskiĭ 14. If X is completely regular and metacompact, is X the image of a paracompact space under a compact open map?

Heath 15. Is every linear-order-topology space with a point countable base quasi-metrizable? (No if Souslin line.)

Lutzer 16. Is a weak θ-refinable, collectionwise normal space paracompact? (No if (MA + $\neg$CH).)

Smith 17. Are compact (or paracompact Σ) spaces with a $\delta\theta$ base metrizable?

B is a *$\delta\theta$ base* for X if $B = \bigcup_{n\in\omega} B_n$ and $x \in X$ and U a neighborhood of x imply there is an n_x such that $\{V \in B_{n_x} \mid x \in V\}$ is a finite nonempty subset of U.

Nyikos 18. In screenable spaces do normal and collectionwise normal imply countably paracompact?

X *is screenable* if every open cover of X has a refinement $R = \bigcup_{n\in\omega} R_n$ where each R_n is a family of disjoint open sets.

Juhász 19. Suppose that X is a hereditarily Lindelöf space of weight $> c$. Is the number of closed sets in $\{Z \subset X \mid w(Z) \le c\}$ of cardinality $\le c$?

Hodel 20. Does a regular ** p-space (or $w\Delta$-space) have a countable base? For ** use ω_1-compact with a point-countable separating open cover, or hereditarily ccc, or hereditarily ccc with a G_δ-diagonal. (The answer is known to be yes for hereditarily ccc with a point-countable separating open cover for $w\Delta$-spaces.)

Lutzer 21. Let C_X be the set of all bounded real valued continuous functions on X; let T be the sup-norm topology on C_X; and let T' be the topology of pointwise convergence. If $A \subset X$, an extender from A to X is a function $e\colon C_A \to C_X$ such that $e(f)$ extends f to X for all $f \in C_A$; e is linear if $e(f + rg) = e(f) + re(g)$ for all real numbers r.

(a) Is there a continuous in T extender from $\beta N - N$ into βN?

(b) Suppose that for every closed subset A of a Moore space X there is a continuous in T linear extender from A to X. Does X have ccc?

(c) Suppose that for every closed subset A of a first countable space X there is a continuous in T' extender from A to X. Must X be collectionwise Hausdorff?

E. Moore Space Problems

Reed 1. Is there a collectionwise Hausdorff nonnormal Moore space?

R. 2. Is each collectionwise Hausdorff, σ-discrete Moore space metrizable?

R. 3. In $V = L$ is each normal Moore space completable? (If (MA + $\neg$CH) there is a normal Moore space which cannot be embedded in a developable space with the Baire property.)

R. 4. Does every Moore space X have a point-countable separating open cover? (Yes if $|X| \le c$.)

Green 5. Does every noncompact Moore space which is closed in every Moore space in which it is embedded have a dense subset which is countably compact but not compact?

Cook 6. If $G_1, G_2, \cdots$ is a development for a Moore space X and $\overline{G^*_{n+1}(p)} \subset G^*_n(p)$ for all n, is every conditionally compact subset of X compact?

Reed 7. Can every first countable space X with $|X| \le c$ be embedded in a separable, first countable space?

R. 8. Can every Moore space X with $|X| \le c$ be embedded in a separable Moore space?

R. 9. Can every metric space X with $|X| \le c$ be embedded in a pseudo-compact Moore space?

Tall 10. Is the product of two normal Moore spaces normal?

T. 11. Is every para-Lindelöf (countably paracompact, Moore) normal space paracompact?

T. 12. Is a normal, locally compact, metacompact space paracompact?

F. Normality of Product Problems

Przymusinski 1. Is there a (first countable separable) paracompact X such that X^2 is normal but not paracompact? (Yes if (MA + $\neg$CH).)

P. 2. Is there a nonparacompact separable first countable space such that X^ω is perfectly normal? (Yes if (MA + $\neg$CH).)

P. 3. Is there a paracompact, separable, first countable space such that X^ω is normal but not paracompact? (Yes if (MA + $\neg$CH).)

P. 4. Is there a locally compact normal space X and a metric space Y such that $X \times Y$ is not normal? (Yes if a Souslin line.)

Howes 5. Does linearly Lindelöf imply Lindelöf in normal spaces?

X is linearly Lindelöf provided any open cover $\{U_r\}_{r<m}$ of X indexed by ordinals with $U_r \subset U_s$ for all $r < s$ has a countable subcover.

H. 6. Is every normal, finally compact in the sense of complete accumulation point space Lindelöf? (Same problem as 5.)

A space X is fcitsocap provided, for every uncountable regular cardinal m and $Y \subset X$ with $|Y| = m$, there is a point x such that every $|U \cap Y| = m$ for all neighborhoods U of x.

Starbird 7. If X is normal and C is a closed subset of X and $f\colon (C \times I) \cup (X \times \{0\}) \to Y$ is continuous, then can f be extended to $X \times I$ if Y is an ANR (normal)?

An ANR (normal) is an absolute neighborhood retract in every normal space in which it is embedded.

S. 8. Can $X \times Y$ be Dowker without either X or Y being Dowker?

S. 9. Let $N(X)$ be the class of all spaces whose product with X is normal. Is $N(X)$ closed under closed maps for paracompact spaces? for paracompact p-spaces?

A paracompact p-space is a closed subset of a product of a compact and a metric space.

Kunen 10. Suppose that T is compact and that Y is the image of X under a perfect map, X is normal, and $Y \times T$ is normal. Is $X \times T$ normal?

Stone 11. Is the box product of ω_1 copies of $\omega + 1$ normal? paracompact?

Nagami 12. Does $\dim(X \times Y) \leq \dim X + \dim Y$ hold for completely regular spaces?

N. 13. Is the image of a μ-space under a perfect maps always a μ-space?

A *μ-space* is a subset of a countable product of paracompact σ-spaces and a σ-space is the union of countably many closed metric subsets.

Corson 14. Is a Σ-product of metric spaces always normal?

S is a Σ-product if there is a product P and a point $p \in P$ such that S is the set of all points of P which agree with p on all but countably many coordinates.

G. Continua Theory Problems

Erdös 1. Is there a connected set in the plane which meets every vertical line in precisely two points such that every nondegenerate connected subset meets some vertical line in two points?

Bing 2. If P is a pseudoarc and $f: P \to P$ is continuous and fixed on an open set, then is f a homeomorphism?

Erdös 3. Is there a widely connected complete metric space? (X is *widely connected* if each nondegenerate connected subset is dense.)

E. 4. Is there a biconnected space without a dispersion point? (X is *biconnected* if it is not the union of two nondense connected subsets.) (Yes if CH.)

Bing 5. Let S be the pseudo-arc and suppose $f: S \to S$ is a map which is fixed on some nonempty open set. Is f the identity?

Bell 6. Is there a compact subset K of the plane which does not separate the plane and a fixed point free map from K to K?

Borsuk 7. Given $X \subset E^3$ such that X is locally connected and separates E^3 does there exits a fixed point free map from X into X? Can *locally contractible* replace *locally connected*?

Cook 8. Is there a hereditarily indecomposable continuum which contains a copy of every hereditarily indecomposable continuum?

Cook (*Knaster*) 9. Is the pseudo-arc a retract of every hereditarily indecomposable continuum in which it is embedded?

Lelek 10. If X and Y are continua, a map $f: X \to Y$ is *confluent* if, for each subcontinua K of Y and component C of $f^{-1}(K)$, $f(C) = K$. Is the confluent image of a chainable continuum chainable?

L. 11. Does the confluent image of a continuum with span 0 have span 0?

X has span 0 if every subcontinuum Z of $X \times X$ such that $\pi_1 \restriction Z = \pi_2 \restriction Z$ intersects the diagonal.

Cook 12. Suppose that $f_1: X_1 \to Y_1$ is confluent and that $f_2: X_2 \to Y_2$ is confluent. Is $f_1 \times f_2: X_1 \times X_2 \to Y_1 \times Y_2$ confluent?

C. 13. Is every continuum with 0 span chainable?

14. Is there a *hereditarily equivalent* (homeomorphic) to each of its nondegenerate subcontinua) continuum which is not tree like?

H. MAPPING PROBLEMS

DEFINITIONS. A map $f: X \to Y$ is *quotient* if, for all $S \subset Y$, S is closed in Y whenever $f^{-1}(S)$ is closed in X. f is countably *biquotient* if, for each $y \in Y$, every (countable) collection of open sets covering $f^{-1}(y)$ has a finite subcollection whose images cover a neighborhood of y. f is *hereditarily quotient* if f_S is quotient for all $S \subset Y$. f is an *s-map* (*L-map*) if $f^{-1}(y)$ is separable (Lindelöf) for all $y \in Y$. f is compact-covering if every compact subset of Y is the image of some compact subset of X.

A space is *of pointwise countable type* provided each point has a sequence $\{U_n\}_{n\in\omega}$ of neighborhoods such that $\bigcap_{n\in\omega} U_n = C$ is compact and every neighborhood of C contains U_n for some n.

A set $\mathcal{G}$ of subsets of a space X is *equi-Lindelöf* if every open cover $\mathcal{H}$ of X has an open refinement with each $U \in \mathcal{G}$ intersecting at most countably many $V \in \mathcal{H}$.

Michael 1. Is every quotient s-image of a metric space also a compact covering quotient s-image of a metric space?

M. 2. Characterize those spaces Y such that every closed map $f: X \to Y$ is countably biquotient (perhaps in terms of sequences of subsets of Y).

M. 3. If X is the metrizable image of a complete metric space under a k-covering map, does X have a complete metric?

M. 4. Let $f: X \to Y$ be a quotient map and let E be a subset of Y such that $\{f^{-1}(y) \mid y \in E\}$ is equi-Lindelöf in X. Assume also that whenever $\{F_n\}_{n\in\omega}$ is a decreasing sequence of subsets of Y with a common limit point, then there is an $A_n \subset F_n$ with A_n closed such that $\bigcup_{n\in\omega} A_n$ is not closed. Is f_E then biquotient?

Olson 5. Suppose that $f: X \to Y$ is a quotient L-map, X has a point countable base, and Y is of point countable type. Is f then biquotient?

O. 6. Is there a quotient map $f: X \to Y$ with X locally compact and first countable, Y compact, each $f^{-1}(y)$ compact, and f finite to one but not hereditarily quotient?

O. 7. Is there a paracompact X of point countable type which does not admit a perfect map onto a first countable space?

Nagata 8. Is the image of a metric space under a q-closed map a σ-space?

Bibliography

1. G. Cantor, *Ueber eine Eigenschaft des Inbergriffes aller reellen algebraishen Zahlen*, J. Reine Angew. Math. 77 (1874), 258–262.

2. P. Urysohn, *Über die Mächtigkeit zusammenhängender Mengen*, Math. Ann. **94** (1925), 262–295.

3. P. Alexandroff and P. Urysohn, *Une condition nécessaire et suffisante pour qu'une class (L) soit une classe (B)*, C. R. Acad. Sci. Paris 177 (1923), 1274–1276.

4. A. Tychonoff, *Über die topologische Erweiterung von Räumen*, Math. Ann. **102** (1929), 544–561.

5. K. Gödel, *Consistency of the axiom of choice and the generalized continuum hypothesis with the axioms of set theory*, Princeton Univ. Press, Princeton, N. J., 1940.

6. ———, *Über formal unentscheidbare Sätze der Principia Mathematica und verwandter Systeme*, Monatsh. Math. und Phys. **38** (1931), 173–198.

7. J. Dieudonné, *Une généralisation des espaces compacts*, J. Math. Pures Appl. (9) **23** (1944), 65–76. MR 7, 134.

8. A. H. Stone, *Paracompactness and product spaces*, Bull. Amer. Math. Soc. 54 (1948), 977–982. MR **10**, 204.

9. R. H. Bing, *Metrization of topological spaces*, Canad. J. Math. **3** (1951), 175–186. MR **13**, 264.

10. Ju. M. Smirnov, *A necessary and sufficient condition for metrizability of a topological space*, Dokl. Akad. Nauk SSSR 77 (1951), 197–200. (Russian) MR **12**, 845.

11. J.-I. Nagata, *On a necessary and sufficient condition of metrizability*, J. Inst. Polytech. Osaka City Univ. Ser. A Math. **1** (1950), 93–100. MR **13**, 264.

12. R. H. Sorgenfrey, *On the topological product of paracompact spaces*, Bull. Amer. Math. Soc. **53** (1947), 631–632. MR **8**, 594.

13. E. A. Michael, *A note on paracompact spaces*, Proc. Amer. Math. Soc. **4** (1953), 831–838. MR **15**, 144.

14. ———, *Another note on paracompact spaces*, Proc. Amer. Math. Soc. **8** (1957), 822–828. MR **19**, 299.

15. ———, *Yet another note on paracompact spaces*, Proc. Amer. Math. Soc. **10** (1959), 309–314. MR **21** #4406.

16. ———, *The product of a normal space and a metric space need not be normal*, Bull. Amer. Math. Soc. 69 (1963), 375–376. MR 27 #2956.

17. C. H. Dowker *On countably paracompact spaces*, Canad. J. Math. **3** (1951), 219–224. MR **13**, 264.

18. H. Tamano, *Note on paracompactness*, J. Math. Kyoto Univ. **3** (1963), 137–143. MR **28** #5419.

19. P. J. Cohen, *The independence of the continuum hypothesis*. I, II, Proc. Nat. Acad. Sci. U.S.A. **50** (1963), 1143–1148; ibid. **51** (1964), 105–110. MR **28** #1118; #2962.

20. L. Gillman and M. Jerison, *Rings of continuous functions*, University Series in Higher Math., Van Nostrand, Princeton, N. J., 1960. MR **22** #6994.

21. E. Hewitt, *Rings of real-valued continuous functions*. I, Trans. Amer. Math. Soc. **64** (1948), 45–99. MR **10**, 126.

22. I. Juhász, *Cardinal functions in topology*, Math. Centre Tracts, 34, Mathematical Centre, Amsterdam, 1971.

23. P. Erdös and A. Tarski, *On families of mutually exclusive sets*, Ann. of Math. (2) **44** (1943), 315–329. MR **4**, 269.

24. M. Reed and P. Zenor, *Preimages of metric spaces*, Bull. Amer. Math. Soc. **80** (1974), 879–880.

25. W. Sierpiński, *Sur un problème de la théorie des relations*, Ann. Scuola Norm. Sup. Pisa (2) **2** (1933), 285–287.

26. F. P. Ramsey, *On a problem of formal logic*, Proc. London Math. Soc. **30** (1930), 264–286.

27. B. Dushnik and E. W. Miller, *Partially ordered sets*, Amer. J. Math. **63** (1941), 600–610. MR **3**, 73.

28. P. Erdös and R. Rado, *Intersection theorems for systems of sets*, J. London Math. Soc. **35** (1960), 85–90. MR **22** #2554.

29. A. Hajnal and I. Juhász, *Discrete subspaces of topological spaces*, Nederl. Akad. Wetensch. Proc. Ser. A **70** = Indag. Math. **29** (1967), 343–356. MR **37** #4769.

30. A. V. Arhangel'skiĭ, *On the cardinality of bicompacta satisfying the first axiom of countability*, Dokl. Akad. Nauk SSSR **187** (1969), 967–970 = Soviet. Math. Dokl. **10** (1969), 951–955. MR **40** #4922.

31. N. Noble and M. Ulmer, *Factoring functions on Cartesian products*, Trans. Amer. Math. Soc. **163** (1972), 329–339. MR **44** #5917.

32. P. S. Urysohn, *Works on topology and other fields of mathematics*, Vol. II, GITTL, Moscow, 1951. (Russian) MR **14**, 122.

33. W. W. Comfort, *A survey or cardinal invariants*, General Topology and Appl. **1** (1971), no. 2, 163–199. MR **44** #7510.

34. M. E. Rudin, *Countable paracompactness and Souslin's problem*, Canad. J. Math. **7** (1955), 543–547. MR **17**, 391.

35. G. Kurepa, *Ensembles linéaires et une classe de tableaux ramifiés*, Publ. Math. Univ. Belgrade **6** (1937), 129–160.

36. D. A. Martin and R. M. Solovay, *Internal Cohen extensions*, Ann. Math. Logic **2** (1970), no. 2, 143–178. MR **42** #5787.

37. I. Juhász, *Martin's axiom solves Ponomarev's problem*, Bull. Acad. Polon. Sci. Sér. Sci. Math. Astronom. Phys. **18** (1970), 71–74. MR **41** #9196.

38. T. Jech, *Non-provability of Souslin's hypothesis*, Comment. Math. Univ. Carolinae **8** (1967), 291–305. MR **35** #6564.

39. W. Fleissner, *When is Jones' space normal?*, (to appear).

40. F. B. Jones, *Concerning normal and completely normal spaces*, Bull. Amer. Math. Soc. **43** (1937), 671–677.

41. E. Specker, *Sur un problème de Sikorski*, Colloq. Math. **2** (1949), 9–12. MR **12**, 597.

42. F. B. Jones, *On certain well-ordered monotone collections of sets*, J. Elisha Mitchell Sci. Soc. **69** (1953), 30–34. MR **15**, 18; 1139.

43. M. E. Rudin, *Souslin's conjecture*, Amer. Math. Monthly **76** (1969), 1113–1119. MR **42** #5212.

44. M. Souslin, *Problème* 3, Fund. Math. **1** (1920), 223.

45. E. W. Miller, *A note on Souslin's problem*, Amer. J. Math. **65** (1943), 673–678. MR **5**, 173.

46. R. B. Jensen, *The fine structure of the constructible hierarchy. With a section by Jack Silver*, Ann. Math. Logic **4** (1972), 229–308; Erratum, ibid. **4** (1972), 443. MR **46** #8834.

47. M. E. Rudin, *A normal space X for which $X \times I$ is not normal*, Fund. Math. **78** (1971/72), no. 2, 179–186. MR **45** #2660.

48. ———, *Souslin trees and Dowker spaces*, Coll. Math. Soc. János Bolyai (8) Topics in Topology, Keszthely, Hungary (1972), 557–562.

49. ———, *A separable Dowker space*, Symposia Mathematica, Instituto Nazionale di Alta Mathematica, 1973 (to appear).

50. R. M. Solovay and S. Tennenbaum, *Iterated Cohen extensions and Souslin's problem*, Ann. of Math. (2) **94** (1971), 201–245.

51. A. M. Gleason, *Projective topological spaces*, Illinois J. Math. **2** (1958), 482–489. MR **22** #12509.

52. D. Booth, *Countably indexed ultrafilters*, Ph.D. Thesis, University of Wisconsin, Madison, Wis., 1969.

53. T. Przymusiński, *A Lindelöf space X such that X^2 is normal but not paracompact*, Fund. Math. **78** (1973), 291–296.

54. F. Tall, *Set-theoretic consistency results and topological theorems concerning the normal Moore space conjecture and related problems*, Ph.D. Thesis, University of Wisconsin, Madison, Wis., 1969.

55. T. Przymusiński and F. Tall, *The undecidability of the existence of a non-separable normal Moore space satisfying the countable chain condition*, Fund. Math. (to appear).

56. F. Tall, *An alternative to the continuum hypothesis and its uses in general topology* (to appear).

57. W. Fleissner, *A normal collectionwise Hausdorff not collectionwise normal space* (to appear).

58. R. H. Bing, *A translation of the normal Moore space conjecture*, Proc. Amer. Math. Soc. **16** (1965), 612–619. MR **31** #6201.

59. R. D. Traylor, *On normality, pointwise paracompactness, and the metrization question*, Topology Conference (Arizona State Univ., Tempe, Ariz., 1967), Arizona State Univ., Tempe, Ariz., 1968, pp. 286–289.

60. F. B. Jones, *Remarks on the normal Moore space metrization problem*, Topology Seminar Wisconsin, 1965, pp. 115–119.

61. A. Hajnal and I. Juhász, *On hereditarily α-Lindelöf and hereditarily α-separable spaces*, Ann. Univ. Sci. Budapest. Eötvös Sect. Math. **11** (1968), 115–124. MR **39** #2124.

62. J. Roitman, *Hereditary topological properties*, Thesis, University of California, Berkeley, Calif., 1974.

63. F. B. Jones, *Hereditarily separable, non-completely regular spaces*, Proceedings of the Blacksburg Virginia Topological Conference, March 1973.

64. M. E. Rudin, *A normal hereditarily separable non-Lindelöf space*, Illinois J. Math. **16** (1972), 621–626. MR **46** #8173.

65. ———, *A non-normal hereditarily separable space*, Illinois. J. Math. **18** (1974), 481–483.

66. A. Hajnal and I. Juhász, *A consistency result concerning hereditarily α-separable spaces*, Indag. Math. **35** (1973), 301–307.

67. ———, *Two consistency results in topology*, Bull. Amer. Math. Soc. **78** (1972), 711. MR **47** #4196.

68. ———, *On first-countable non Lindelöf S spaces*, Math. Inst. Hungar. Acad. Sci., 1973 (preprint).

69. A. J. Ostaszewski, *On countably compact, perfectly normal spaces*, J. London Math. Soc. (to appear).

70. R. B. Jensen, *The fine structure of the constructible universe*, Ann. Math. Logic **4** (1972), 229–308.

71. K. J. Devlin, *Aspects of constructibility*, Lecture Notes in Math., vol. 354, Springer-Verlag, Berlin and New York, 1973.

72. W. Fleissner, *When normal implies collectionwise Hausdorff*, Thesis, University of California, Berkeley, Calif., 1974.

73. W. Rudin, *Homogeneity problems in the theory of Čech compactifications*, Duke Math. J. **23** (1956), 409–419. MR **18**, 324.

74. Z. Frolík, *Sums of ultrafilters*, Bull. Amer. Math. Soc. **73** (1967), 87–91. MR **34** #3525.

75. J. Ketonen, *On the existence of P-points* (unpublished).

76. N. M. Warren, *Properties of Stone-Čech compactifications of discrete spaces*, Proc. Amer. Math. Soc. **33** (1972), 599–606. MR **45** #1123.

77. K. Kunen, *On the compactification of the integers*, Notices Amer. Math. Soc. 17 (1970), 299.

78. E. Čech, *On bicompact spaces*, Ann. of Math. 38 (1937), 823–844.

79. M. E. Rudin, *Partial orders on the types in βN*, Trans. Amer. Math. Soc. 155 (1971), 353–362. MR 42 #8459.

80. A. Blass, *The Rudin-Keisler ordering of P-points*, Trans. Amer. Math. Soc. 179 (1973), 145–166.

81. R. Hodel, *Extensions of metrization theorems to higher cardinality*, Fund. Math. (to appear).

82. W. W. Comfort and S. Negrepontis, *Homeomorphs of three subspaces of $\beta N \setminus N$*, Math. Z. 107 (1968), 53–58. MR 38 #2739.

83. R. Hodel, *Spaces defined by sequences of open covers which guarantee that certain sequences have cluster points*, Duke Math. J. 39 (1972), 253–263. MR 45 #2657.

84. M. Katětov, *On real-valued functions in topological spaces*, Fund. Math. 38 (1951), 85–91. MR 14, 304.

85. M. Starbird and M. E. Rudin, *Products with a metric factor*, General Topology and Appl. (to appear).

86. M. Starbird, *The normality of products with a compact or a metric factor*, Thesis, University of Wisconsin, Madison, Wis., 1974.

87. M. E. Rudin, *Products with one compact factor*, General Topology and Appl. (to appear).

88. K. Morita, *Note on paracompactness*, Proc. Japan Acad. 37 (1961), 1–3. MR 25 #3502.

89. ———, *On generalizations of Borsuk's homotopy extension theorem*, Fund. Math. (to appear).

90. K. Borsuk, *Sur les prolongements des transformations continues*, Fund. Math. 28 (1937), 99–110.

91. C. J. Knight, *Box topologies*, Quart. J. Math. Oxford Ser. (2) 15 (1964), 41–54. MR 28 #3398.

92. M. E. Rudin, *The box product of countably many compact metric spaces*, General Topology and Appl. 2 (1972), 293–298.

93. ———, *Countable box products of ordinals*, Trans. Amer. Math. Soc. 192 (1974), 121–128.

94. P. Erdös and M. E. Rudin, *A non-normal box product*, Coll. Math. Soc. János Bolyai (10) Keszteley, Hungary, 1973.

95. Ken Kunen, *Box products of compact spaces* (to appear).

96. ———, *On normality of box products of ordinals* (to appear).

97. ———, *Some comments on box products*, Coll. Math. Soc. János Bolyai (10) Keszteley, Hungary, 1973.

98. E. K. Van Douwen, *The box product of countably many metrizable spaces need not be normal*, Fund. Math. (to appear).

99. J. E. Vaughan, *Non-normal products of ω_μ-metrizable spaces* (to appear).

100. R. Engelking and M. Karłowicz, *Some theorems of set theory and their topological consequences*, Fund. Math. 57 (1965), 275–285; Corollary on p. 282. MR 33 #4880.

101. W. W. Comfort and S. Negrepontis, *The theory of ultrafilters*, Die Grundlehren der math. Wissenschaften, Band 211, Springer-Verlag, Berlin and New York, 1974.

102. T. J. Jech, *Trees*, J. Symbolic Logic 36 (1971), 1–14. MR 44 #1560.

103. R. W. Heath, *Arc-wise connectedness in semi-metric spaces*, Pacific J. Math. 12 (1962), 1301–1319. MR 29 #4032.

104. ———, *Screenability, pointwise paracompactness, and metrization of Moore spaces*, Canad. J. Math. 16 (1964), 763–770. MR 29 #4033.

105. S. Heckler, *On a ubiquitous cardinal*, Proc. Amer. Math. Soc. (to appear).

The following papers are not refered to in the text; but anyone interested in set theoretic topology should be acquainted with the important work being done by the Russian school. The papers listed here are a modest selection made by J. M. Smirnov of some of the excellent papers related to the problems discussed in these notes.

106. A. V. Arhangel'skiĭ, *Suslin number and power. Characters of points in sequential bicompacta*, Dokl. Akad. Nauk SSSR 192 (1970), 255–258 = Soviet Math. Dokl. 11 (1970) no. 3, 597–601. MR 41 #7607.

107. ———, *Bicompacta that satisfy the Suslin condition hereditarily. Tightness and free sequences*, Dokl. Akad. Nauk SSSR 199 (1971), 1227–1230 = Soviet Math. Dokl. 12 (1971) no. 4, 1253–1257. MR 44 #5914.

108. ———, *On cardinal invariants*, Proc. Third Prague Topology Symp., 1971, 37–46.

109. ———, *There is no "naive" example of a non-separable sequential bicompactum with the Suslin property*, Dokl. Akad. Nauk SSSR 203 (1972), 983–985 = Soveit Math. Dokl. 13 (1972) no. 2, 473–476. MR 45 #9286.

110. ———, *The property of paracompactness in the class of perfectly normal, locally bicompact spaces*, Dokl. Akad. Nauk SSSR 203 (1972), 1231–1234 = Soviet Math. Dokl. 13 (1972) no. 2, 517–520.

111. ———, *Frequency spectrum of a topologcial space and the classification of spaces*, Dokl. Akad. Nauk SSSR 206 (1972), 265–268 = Soviet Math. Dokl. 13 (1973) no. 5, 1185–1189.

112. V. I. Ponomarev, *The cardinality of bicompacta satisfying the first axiom of countability*, Dokl. Akad. Nauk SSSR 196 (1971), 296–298 = Soviet Math. Dokl. 12 (1971) no. 1, 121–124. MR 43 #1122.

113. B. A. Efimov, *Dyadic bicompacta*, Trudy Moskov. Mat. Obšč. 14 (1965), 211–247. MR 34 #1979.

114. ———, *On the power of Hausdorff spaces*, Dokl. Akad. Nauk SSSR 164 (1965), 967–970 = Soviet Math. Dokl. 6 (1965) no. 5, 1315–1318. MR 32 #8301.

115. B. A. Efimov, *On embedding of Stone-Čech compactifications of discrete spaces in bicompacta*, Dokl. Akad. Nauk SSSR **189** (1969), 244–246 = Soviet Math. Dokl. **10** (1969) no. 6, 1391–1394. MR **40** #6505.

116. ———, *Subspaces of dyadic bicompacta*, Dokl. Akad. Nauk SSSR **185** (1969), 987–990 = Soviet Math. Dokl. **10** (1969) no. 2, 453–456. MR **39** #3459.

117. ———, *Extremally disconnected compact spaces and absolutes*, Trudy Moskov. Mat. Obšč. **23** (1970), 243–285.

118. V. V. Fedorčhuk, *Bicompacta in which every infinite closed subset is n-dimensional*, Mat. Sb. (1975) (to appear).

119. ———, *Bicompacta without canonically correct subsets*, Dokl. Akad. Nauk SSSR **215** (1974) (to appear).

120. V. I. Malyhin, *On countable spaces having no bicompactification of countable tightness*, Dokl. Akad. Nauk SSSR **206** (1972), 1293–1296 = Soviet Math. Dokl. **13** (1972) no. 5, 1407–1411.

121. ———, *Nonnormality of some subspaces of βX, where X is a discrete space*, Dokl. Akad. Nauk SSSR **211** (1973), 781–783 = Soviet Math. Dokl. **14** (1973) no. 4, 1112–1115.

122. ———, *On dispersed and maximal spaces*, Dokl. Akad. Nauk SSSR **214** (1974) (to appear).

123. N. A. Šanin, *On the product of topological spaces*, Trudy Mat. Inst. Steklov. **24** (1948). MR **10**, 287.

124. B. È. Šapirovskiĭ, *Discrete subspaces of topological spaces. Weight, tightness and Suslin number*, Dokl. Akad. Nauk SSSR **202** (1972), 779–782 = Soviet Math. Dokl. **13** (1972) no. 1, 215–219. MR **45** #1100.

125. ———, *On the density of topological spaces*, Dokl. Akad. Nauk SSSR **206** (1972), 559–562 = Soviet Math. Dokl. **13** (1972) no. 5, 1271–1275.

126. ———, *Canonical sets and character. Density and weight in compact spaces*, Dokl. Akad. Nauk SSSR **218** (1974), 58–61 = Soviet Math. Dokl. **15** (1974) (to appear).

127. ———, *On spaces with Souslin's and Šanin's property*, Mat. Zametki **15** (1974) no. 2.

128. V. I. Malyhin and B. È. Šapirovskiĭ, *Martin's axiom and properties of topological spaces*, Dokl. Akad. Nauk SSSR **213** (1973), 532–535 = Soviet Math. Dokl. **14** (1973) no. 6, 1746–1751.

129. G. P. Amirdžanov, *Density, Souslin number, and tightness of topological spaces*, Dokl. Akad. Nauk SSSR **209** (1973), 265–268 = Soviet Math. Dokl. **14** (1973) no. no. 2, 350–354.

130. G. P. Amirdžanov and B. È. Šapirovskiĭ, *On everywhere dense subsets of topological spaces*, Dokl. Akad. Nauk SSSR **214** (1974), 249–252 = Soviet Math. Dokl. **15** (1974) no. 1, 87–92.